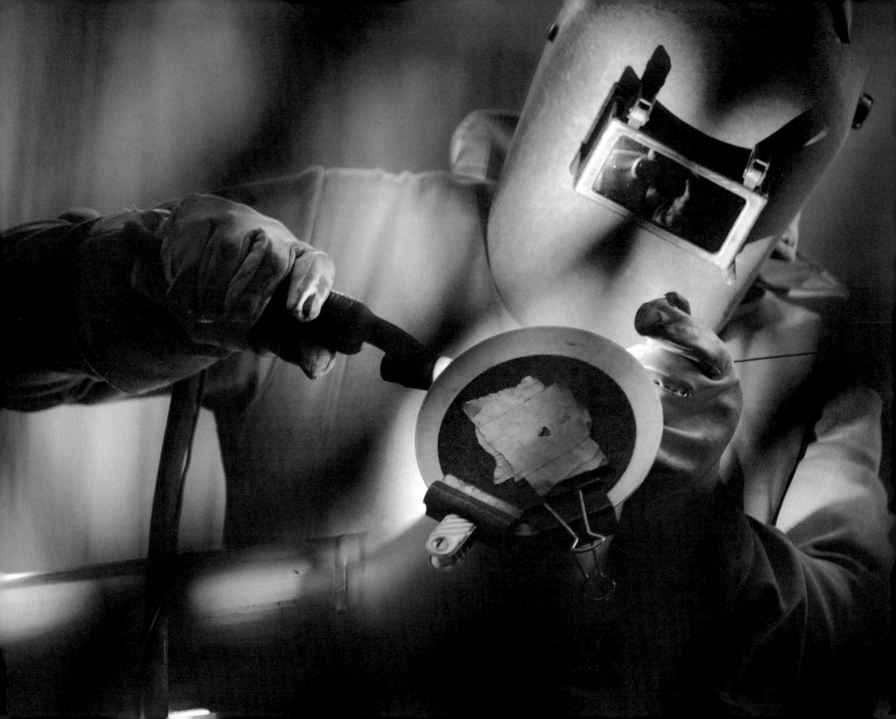

GLENGLASSAUGH
A Distillery Reborn

IAN BUXTON

PUBLISHED BY THE ANGELS' SHARE
IN ASSOCIATION WITH
THE GLENGLASSAUGH DISTILLERY CO LTD

IS AN IMPRINT OF
NEIL WILSON PUBLISHING LTD
G/2 19 NETHERTON AVENUE
GLASGOW G13 1BQ

WWW.NWP.CO.UK

ISBN 978-1-906476-10-6 (PAPERBACK)
ISBN 978-1-906476-13-7 (HARDBACK)

FIRST PUBLISHED IN JULY 2010

COPYRIGHT © 2010, IAN BUXTON.

FOREWORD COPYRIGHT © 2010, ALEX SALMOND

ALL RIGHTS RESERVED
No part of this publication may be reproduced, stored in an information retrieval system, or transmitted, in any form or by any means, electronic, mechanical, photocopying, recording or otherwise, without the prior written permission of the publisher.

A CATALOGUE RECORD FOR THIS BOOK IS AVAILABLE FROM THE BRITISH LIBRARY

IAN BUXTON HEREBY ASSERTS HIS MORAL RIGHT TO BE IDENTIFIED AS THE AUTHOR OF THIS WORK
IN ACCORDANCE WITH THE COPYRIGHT, DESIGN & PATENTS ACT, 1988

DESIGNED BY JULES AKEL

PRINTED & BOUND BY
1010 PRINTING INTERNATIONAL LTD., NORTH POINT, HONG KONG

CONTENTS

Foreword … i

Introduction … iv

1. The Foundation Years … 1

2. Enter 'The Highland' … 21

3. Clinging on — Rare & Unusual Bottlings … 48

4. Renaissance … 60

5. Something Old, Something New — The Current Expressions … 85

6. A Run Through Glenglassaugh With Alfred Barnard … 100

Credits & Bibliography … 115

Foreword

THE RIGHT HONOURABLE ALEX SALMOND MSP MP

FIRST MINISTER OF SCOTLAND

Whisky is one of Scotland's finest and most valuable exports and our many varied malts and blends are savoured and appreciated around the world. Scotch Whisky is an iconic symbol at times of celebration both at home and abroad—from toasting the Bard at a Burns Supper to a dram at Hogmanay. For over 130 years Glenglassaugh distillery has played its part in spreading the pleasure of whisky across the globe.

As the local Member of Parliament it was an honour to officially re-open the *Glenglassaugh* distillery in November 2008 and it was a particularly uplifting experience to see such an iconic distillery brought back to life. Over its long history *Glenglassaugh* has seen several periods of operation interspersed with silent mothballed interludes and it is wonderful the distillery is committed to using traditional equipment and methods to once again make its unique whisky.

A Distillery Reborn offers a fascinating insight into the history of *Glenglassaugh* and its substantial contribution to the local community over the years. We should raise a glass to its continued success.

Introduction

IAN BUXTON

This is a rather unusual whisky book. Neither a completely independent history, nor an insider's account, it may be best thought of as 'semi-detached'. To understand that curious description requires a brief history of my involvement with Glenglassaugh. As will become clear elsewhere, the distillery was silent and mothballed when purchased by its new owners. Consequently, it had been all but forgotten by the world of whisky and certainly when I was first approached about Glenglassaugh I had never visited it nor, to the best of my recollection, had I even tasted the spirit.

I can place that approach exactly: it was December 20th 2007 at *The Mash Tun*, a well-known whisky bar with rooms in Aberlour, and the hour of that meeting was 9.30 a.m. I was due to meet Stuart Nickerson, an independent consultant working in the whisky industry, who I knew slightly from his previous employment as Distilleries Director of William Grant and Sons (the owners of Glenfiddich and The Balvenie, amongst many other things).

Nickerson had left Grants some years previously and was working as a 'consultant'. That can mean many things, as I know myself.

I once saw a cartoon in which a middle-aged executive stared at an empty desk, evidently a home office, with a thought bubble above his head reading 'I am not unemployed, I am a consultant'. It is a phenomenon that many will recognise and could, not entirely without justification, have been applied to the early days of my own self-employment.

I was vague as to the purpose of our meeting. In the brief call setting it up, Stuart had referred to a syndicate of investors who were proposing to buy a distillery and needed marketing expertise. He had been acting for them and now a deal was imminent—was I interested?

To be entirely honest, I was sceptical. Calls of this nature were not unusual in 2007 and 2008 as the whisky boom gathered pace. There were a number of more or less speculative schemes to purchase and re-open redundant distilleries or to build new ones, some in far-flung locations. Few of these plans came to anything and, along with a number of other whisky writers and commentators, I had become both suspicious and worried about some of the claims being made by some of the promoters. It was rapidly becoming apparent to many within the industry that small 'investors' were at risk and money was going to be lost, though it was far from clear what could be done about this. Whilst the schemes being floated were highly speculative and evidently more a gamble than an investment, they were not criminal or fraudulent—*caveat emptor* seemed to be the motto.

However, apart from a general concern about this, I was alarmed by the potential for the reputational damage to Scotch whisky if and when such schemes failed, as I believed they would. Accordingly, I was in no great hurry to associate my name with them, though tempted by some of the fees on offer. So I approached the meeting with some trepidation.

However, as the details emerged, I relaxed and began to feel more curious. Stuart explained that he had been working for his investors—at that time no details were provided—for nearly two years. Two deals to purchase a distillery had come close to completion but fallen through at the last minute due, he claimed, to a change of heart by the different vendors.

He also provided some details, in confidence, of the proposed deal to buy *Glenglassaugh* which he assured me would be completed without incident. As I learned more so I felt happier: the distillery being sold (no name was mentioned at this stage) was owned by Highland Distillers.

Knowing something of that company I knew that they would have done their own due diligence into the prospective purchasers and satisfied themselves not only that the funds were forthcoming, but that the new owners understood the business they were getting into and that they had the resources and commitment to see the project through. For it is one thing to buy a distillery, it is altogether another to restore it to working condition and then operate it for the length of time necessary for mature whisky to be available for sale.

If Highland Distillers were happy, I reasoned, so was I. As private and almost secretive as they are at a corporate level this is a company that guards its justifiably high reputation with great jealousy and would be at pains not to sully this in any way. If they were prepared to sell they must have been reassured about the purchasers and entirely happy that this sale could not reflect badly on them. Better by far to keep the distillery mothballed than to pocket the sale price, however high. No amount of money would compensate them for even a slight loss of reputation so I knew the deal must have been sanctioned only after the most careful consideration. Ultimately Highland Distillers is owned by a charitable trust that is rooted in the highest of traditional Scottish values—reserved, discreet and conservative but highly admirable.

And then there was the fact that this was Stuart Nickerson that I was speaking to. As a distiller, his reputation went before him. With experience at Highland Park, Glenrothes, Glenfiddich, The Balvenie and Grants' giant grain whisky operation in Girvan (and, as it turned out, some knowledge of *Glenglassaugh* also) he had forgotten more about making whisky than I would ever know. And I considered the man himself: Stuart is by nature almost a caricature of the canny Scot but passionate about what he does and, as I have come to learn, a man of his word. The fact of his long-term involvement and his decision to give up his consultancy practice to return to a full-time management role both intrigued and reassured me.

As he explained more, so my interest grew. We met several times before the deal was concluded and our own arrangements finalised. I learnt that his investors were a Dutch-registered investment group with interests in the energy market, most notably in Russia and the former Soviet states. I learnt that they were fully committed to bringing the distillery back to work and to producing the highest quality of spirit. I met some of their senior executives and liked and respected them. I learnt that Stuart had managed

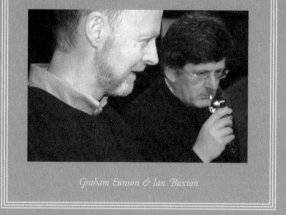
Graham Eunson & Ian Buxton

to lure Graham Eunson, a quietly-spoken Orcadian, from Glenmorangie to be his new distillery manager and my interest grew. I learnt that along with the distillery buildings and the brand name Highland Distillers were selling all their remaining stocks of mature *Glenglassaugh*; as I tasted it interest turned to excitement.

What was not to like? It was clear that for most of the first year Stuart, Graham and their team were to be fully engaged in refurbishing the distillery and getting it running; my task was to help develop a marketing and brand strategy and begin to execute it while the distillers worked out what kind of an animal they were trying to tame at *Glenglassaugh*.

So I started work in March 2008, as soon as the distillery's purchase was completed.

My role was that of an interim, part-time Director of Marketing. At that stage *Glenglassaugh* did not require, nor could it justify, a full time marketing resource. The priority was on fixing the distillery; while getting some of the mature stock to market was clearly important it took second place and, in any event, there was not enough mature *Glenglassaugh* to support the kind of marketing infrastructure and expenditure that characterises many of its competitors.

So I was *of* the team, but not wholly *in* it. Sometimes this resulted in frustration, but more often it provided a helpful detachment (or so I firmly believe). It was always planned that my involvement would run down. Once the key elements of the strategy were in place the requirement was for execution and delivery of the plans. This is both a more labour intensive and time consuming role than I was able to undertake given my other commitments, and in any event not the proper responsibility of a consultant.

But it will always be a source of pride and

pleasure that I was able to get on site within hours of the purchase being finalised; that I played a major role in the re-opening celebrations led by Scotland's First Minister; that I was in the still room when the very first new spirit in more than twenty years ran at *Glenglassaugh* and that I can point to the new brand identity, packaging and products and say 'I did that' (well, bits of it, anyway).

Subsequently we agreed that I would write this account of the distillery and its re-opening. As you will appreciate from this recital of the circumstances it is a partisan account: this is partly the distillery's story (they have supported the publication of this book) but it is my version of the story also, not a promotional brochure or a piece of marketing literature.

It's too important for that. Stuart Nickerson has supplied much background and historical material but, apart from correcting some facts and advising on technical issues to do with distilling, has given me a free hand to write as balanced an account as I am able, and for this alone he deserves both thanks and respect.

Publication of the book marks a further diminution of my role and this is therefore a bittersweet moment. But I am confident that *Glenglassaugh* will go on from strength to renewed strength.

This is a distillery reborn. This is its story.

Ian Buxton
PITLOCHRY, JANUARY 2010

CHAPTER I

The Foundation Years

'It is but the truth when we say that in order to write anything like a memoir of Col. Moir would be to write a history of every movement that has taken place in town and district for half a century back, for during all those years there has scarcely a movement of any note taken place of which he was not the originator, guiding spirit, or at least a staunch advocate.'

If the *Banffshire Reporter* of 5th October 1887 may be believed, the pleasant seaside town of Portsoy has much cause to be grateful to Colonel James Moir for his good works.

During his fifty years of prominence, Col. Moir established several successful businesses in fields as diverse as banking; the importation of Peruvian guano; fishing, agriculture and shipping. Though not a regular soldier he served energetically in the local Volunteer and was gazetted Lieutenant Colonel of the Banffshire Battalion of Volunteer Artillery, retiring in 1873 but retaining the rank of Honorary Colonel.

On his arrival in Portsoy in 1834, Moir opened a general store in the High Street, and traded as a wine and spirit merchant, seedsman and manure merchant (hence the interest in guano), and ironmonger. His interests grew rapidly and, as agent for the North of Scotland Bank he was at the heart of commercial activity in the town.

He took an active and prominent part in extending the Strathisla Railway to Portsoy in 1859, but not content with giving Portsoy railway communications southwards, he undertook the additional task of securing a Parliamentary Bill for the construction of a line westwards along the coast, in which effort he was successful. As Chairman and principal shareholder of the Portsoy Gas Company he brought the first mains gas supply to the town and in 1868, to promote the local fisheries he was instrumental in linking Portsoy, Buckie and Portgordon by a 'sixpenny telegraph'—the wire being later sold to the Government, no doubt at a handsome profit.

With his interests in shipping and fishing Moir

was an enthusiastic participant in the Portsoy Harbour Committee and actively interested in the harbour's reconstruction, bringing great benefits to the town. In acknowledging his prodigious physical energy the author of the *Banffshire Reporter* obituary noted, perhaps a trifle enviously, that *'He was a great pedestrian in his day, and for years is said to have walked between Portsoy and Turriff regularly every Saturday'*. That is a distance of some 17 miles, though the mode of the Colonel's return transport is not recorded.

Finally, in July 1887, shortly before his death, this worthy citizen and model of Victorian commercial acumen and drive gifted land and property in Seafield Street as a site for the new Town Hall. Lacking issue from either of his marriages this archetypal philanthropy was perhaps intended as his memorial.

None of this concerns us, however, other than to note his considerable, one might even say ubiquitous, presence in local affairs. Clearly, if James Moir determined that a distillery should be built, a distillery would be built—and a fine one at that. And so, in 1875, the *Glenglassaugh Distillery Company* came to pass, duly if imaginatively eulogised some twelve years later by the *Banffshire Reporter* as *'a concern which has now assumed gigantic dimensions'*.

However it was not, it must be confessed, an entirely original idea. There had previously been a distillery right in the heart of Portsoy. The Portsoy Distillery Co. is first heard of in the *Edinburgh Advertiser* of 1800. It traded from premises in Low Street, just off the harbour, where by 1830 it was known as Burnside Distillery. It appears to have ceased operations at some time after 1837, so Moir would have been aware of the distillery during his early years of trading as a general merchant and no doubt he sold its products to local customers. Perhaps as a banker he was aware of the reasons for its demise and deplored its impact on the local economy; as a merchant he surely regretted the loss of business.

Whatever its history however, today no trace remains of Burnside. However, in Pigot & Co's Banffshire trade directory for 1825 a William Morrison is described as 'Conductor' of the Portsoy Distillery.

We cannot be certain, but it seems probable that this is the same Morrison who will later feature in the *Glenglassaugh* story as founding partner. In the absence of Burnside, presumably Moir obtained supplies from the nearby Banff Distillery at Inverboyndie, which was demolished as late as 1983.

Around 1834 he married into the family of the Morrisons of Turriff, described simply as 'merchants' and on the death of Peter Morrison came eventually to acquire their business. It was with the Morrison family that he was to promote construction of the distillery. The partners in the *Glenglassaugh Distillery Company* are shown in contemporary documents to be Moir, his two nephews Alexander and William Morrison and Thomas Wilson, described as a coppersmith of Portsoy, though their respective shares are not recorded.

This was a propitious moment in the history of Scotch whisky, a fact of which a shrewd businessman such as Moir would have been keenly aware, even without his interest in the wine and spirit trade. With the advent of blending and the decline of brandy production, Scotch whisky was in the ascendancy even as Irish whiskey, its great competitor, began to falter. A number of notable distilleries were founded around this time, including Cragganmore (1869), Balblair (1872), Glenrothes (1878), Bunnahabhain and Bruichladdich (both 1881).

In fact *Glenglassaugh* slightly anticipated the boom in distillery construction, which accelerated from 1890. In the following decade, a remarkable forty-one new distilleries were opened in Scotland, a pace of development which has never been equalled, even if some of them were short-lived.

To that extent, the decision to open *Glenglassaugh* was a far-sighted one.

Together with the adjacent Craigmills Farm, which provided barley for the distillery, some 17 acres of ground were acquired on the Glassaugh estate at the foot of the Fordyce Burn, round the headland at the east of the town. The site is dominated by the imposing remains of the Sandend

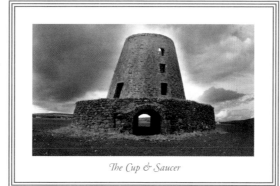
The Cup & Saucer

windmill, today known locally as the 'cup and saucer'. Though not linked to the distillery its proximity often attracts attention, and its story is an interesting one.

According to the *1791–99 Statistical Account of Scotland*, the site originally housed a substantial prehistoric burial mound some fourteen feet high and sixty feet across, covered with turf. When opened it was found to contain a stone coffin and well-preserved bones, presumed those of a chieftain together with a deer's horn, symbolic of a hunter. It has been speculated that stone from the burial mound was employed in the construction of the windmill, in which case history will record another stain on the reputation of James Abercromby who was responsible for erecting this dramatic structure.

Abercromby had recently returned from North America to his Scottish estates somewhat in disgrace following his catastrophic command at the Battle of Ticonderoga (1758).

There he led an army of some 16,000 troops against a much smaller, but well entrenched combined French-Indian enemy force estimated to be some 4,000 strong. British casualties were at least 2,000, substantially heavier than the losses inflicted on the French and Indians, with the Black Watch taking particularly heavy losses. Abercromby retreated in disarray and confusion.

Describing his leadership at Ticonderoga the historian Lawrence Gipson has suggested that *'no military campaign was ever launched on American soil that involved a greater number of errors of judgement on the part of those in positions of responsibility'*.

Contemporary accounts have his officers and troops describing Abercromby in rather more trenchant terms but despite this disaster Abercromby was protected by his political connections. Though he was withdrawn from the North American campaign and never again commanded troops in action, he reached the rank of full General in 1772. Perhaps it was in an attempt to forget his disastrous foray at Ticonderoga that Abercromby turned to building works and the improvement of his Glassaugh estate.

In August 1761 he wrote to his eldest daughter, then resident at their house in London's fashionable Golden Square, complaining of high winds which had threatened *'the pompon of the windmill, which was only set up yesterday'*. This presumably refers to the movable wooden structure atop the tower, now lost, which housed the gears connecting the sails to the driveshaft for the millstones.

Later that year he entertained Lady Lessendrum and her two daughters to a trip to the mill, where they enjoyed two plates of apricots and plums, suggesting that the windmill was still something of a local novelty. Its working life appears to have ended at some time in the early 19th Century when steam power rendered it redundant.

By 1887, when Alfred Barnard first visited the distillery, the mill's upper floors were apparently in use as storage for the distillery. With his distinctive love of idiosyncratic detail and local colour Barnard relates that a walk of some ten minutes *'brought us to the ruins of an ancient mill, built over a lofty brick archway, through which we passed to the Distillery, close by'*.

A track may well have run through the mill, the arched entrances of which could admit a pony and trap, but it was not the distillery's main entrance. However, with that account of the highly distinctive landmark that is the 'cup and saucer' to one side, we can return to *Glenglassaugh*'s history.

The *Glenglassaugh Distillery Company*'s lease permitted them to extract water from the Fordyce Burn—considered excellent for distilling and used, according to local tradition, by several illicit distillers in the years before 1823. Water, which as we shall come to see plays such a pivotal role in *Glenglassaugh*'s history, was thought to hold enormous significance for spirit quality before modern science rather relegated its importance below that of fermentation chemistry, still design and, above all, wood management. For all that, however, Moir and his partners had given considerable thought to the location of their new project and a sound and reliable source of water was regarded as of paramount significance.

They were allowed to cut peats from the moss of Auchinderrom for use in the distillery at a cost of £1 'per spade's casting'. This was a consideration almost as important as the water supply. As the *Portsoy Advertiser* reported in its issue of 12th March 1875 as the distillery neared completion, '*Glenglassaugh Whisky will be the real "peat reek", as peats are largely used in drying the malt*'. All in all, this was thought to be an excellent location in the midst of a '*fine corn providing district*' with '*splendid barley crops*'.

Barnard too, in 1887, drew attention to the '*five large peat stacks, containing upwards of 400 tons*'.

Peat, of course, was hugely important in Scotch whisky production in the nineteenth century and, as late as 1930, no less an authority than Aeneas MacDonald could write, '*The convenient proximity of a peat bog is an economic necessity for Highland malt distillery*'. Whilst this is no longer the case and the distillery's present new make is unpeated, in Nickerson and Eunson's quest for authenticity in production there are plans to begin trials with both medium and heavily peated malt in the next year. It helps that heavily peated styles are currently greatly demanded, especially by northern European single malt aficionados.

That report in the local newspaper is the main source for information on the original *Glenglassaugh* distillery. From it, we learn that '*the plans for it were drawn by Mr Melvin, Elgin; the contractor for the mason work was Mr A. Barclay; slater and plumber work, Mr Geo. McDonald, Portsoy; plaster work,*

Messrs McIvor and Younie, Cullen and Elgin; Stills, Mr T. Wilson, Portsoy. The cast-iron work is from Mr Johnstone's Foundry, Elgin, with exception of the mashing machine which is from Banff Foundry; and the millwright is Mr Petrie, Ballindalloch'.

The Manager was then a Mr Sellar, though he appears to have been replaced shortly after this by Dugald Mathieson who came with his family from Campbeltown, then an important distilling centre.

The distillery's original capacity is also meticulously detailed for, as the report explains, *'in the course of its erection, the work, like all works of this kind, has excited a good deal of interest'.*

As was normal at the time, the distillery would have resembled a small community of its own, with housing for the Manager, workforce and two Excise officers; extensive stores and substantial buildings associated with malting, all of which was undertaken on site. Glassaugh Station was some ten minutes walk away and, were the workforce in need of any refreshment (in addition to the liberal dramming of 'clearic' which would have taken place) there was an inn at Glassaugh smiddy known then as the Black Jug and later as the Red Lion.

Dugald Mathieson

Great care was taken to exploit the topography and take advantage of the supplies of limestone which could be quarried on site. The distillery ran largely on water power—there was a powerful overshot water wheel 17 feet in diameter and 4 feet broad which drove all the equipment and was arranged so that, in the words of the *Portsoy Advertiser, 'with exception of the manufactured spirits, everything may be said to go down, down and further down by gravitation, so to speak'.*

In fact, given today's concerns with energy

M
WITHERING FLOOR

efficiency and the carbon footprint of any business, the operation of the original *Glenglassaugh* distillery may be considered enviably 'green'. Though yields would inevitably have been lower than today, due in part to the different varieties of barley used, the principles of construction and operation seem remarkably contemporary in terms of their environmental impact, though low operating costs would have been the principal objective of this design.

Though, as we shall see, little remains of the 1875 distillery or warehouses, two large grain lofts survive from this period, testifying to the scale of the operation. The 'Withering Floor' may still be seen, together with a large stone steep in the easternmost malting building, presently unused, though earmarked for long-term restoration. Stuart Nickerson's fond hope is that this may eventually be brought back into use and floor malting once again take place at *Glenglassaugh*.

It is certainly an imposing structure—the original kiln (now removed) was capable of drying 35 quarters (around 7,100kg) of barley in 48 hours. From this, the distillery would have expected to receive approximately 5,330 kilos of malt for milling. At this rate of production we can understand why, on opening, the distillery was said to hold between 2,000 and 3,000 quarters of barley.

Beyond the malting was the malt deposit, the floor of which was 2 feet below the level of the floor of the kiln. It was 30 feet long and 14 feet wide, and capable of containing 300 bushels of prepared malt.

For a detailed description of the lost distillery we must turn again to the trusty *Portsoy Advertiser*, as nothing today survives beyond the Excise cottages, two maltings and the original dunnage warehouse. The works were described in the following terms:

'(The) malt mill is capable of mashing 160 bushels in 18 minutes. In the mashing machine the malt is met and blended with the hot water from two coppers, each of

which has a capacity of 1,500 gallons. From the mashing machine the malt and water rush into the mash tun, which is of cast-iron and is of a large size.

'Another pipe from the coppers enters the mash tun on a level with the floor of it, and may be used in washing it out. From the mash tun the wort or wash is pumped up to the cooler and in the bottom of the mash tun a hatch opens to let the draff descend into carts, which will take it to the cattle sheds. The cooler, which like the mash tun is of cast-iron, is 42 feet by 30.

'Underneath the cooler, and of the same dimensions, is the tun room, in which stand what are called the washbacks, or tuns—four in number, each of which is capable of containing 5,700 gallons. From these wash backs the wash, or wort, runs into the wash charger, which has a capacity of 3,700 gallons.

'From this it goes into the wash still, which will hold 2,500 gallons. From the high head of the wash still the low wines and feints descend into the cooling worm, passing through which, they then run into the low wines and feints receiver, from which they again, at the proper time, descend into the spirit still, which will hold 1,500 gallons, and, passing through the cooling worm of that still, the whisky—for it is whisky then—passes into the spirit receiver, and from it the spirits enter a large solitary covered and closely secured tun which is called the spirit vat.

'This vat has a capacity of 2,000 gallons, and is sufficiently elevated above the floor of the room in which it stands to admit of the spirits flowing from it into the casks. In the same room is a powerful weighing beam for weighing the casks, the quantity of spirits being now calculated by weight. Facing the door of the room in which stands the spirit vat, but with a passage wide enough for a cart to pass between them is the door of the bonded warehouse which is 80 feet long by 30 feet in breadth, and is at present on the lowest level of any of the buildings.

'We may explain that the still room does not range with the other buildings, but forms a wing on the east

Glenglassaugh Distillery wishes you a merry Xmas

35 Published by Alex. Robertson, Fordyce

side. The door of it opens to the east, and is close to the side of the burn, and is but little above the level of it. The tank for the cooling worms stands on the south side of it, and is placed at such a level that a cooling stream of water from the mill lade may be kept constantly flowing through it.'

The cost of the new distillery was some £10,000 (or approximately £700,000 in today's money, based on the R.P.I.) and, according to the *Banffshire Reporter* of 16th April 1875 the first brew or 'browst' took place successfully that month.

All the buildings may be seen in the handsome hand-coloured postcard illustrated here, which was produced by Robertsons of Fordyce in the very late 1890s. The 'cup and saucer' of General Abercromby's mill can be clearly discerned in the background to the right of the card.

Though in all probability photographed some years previously, the card is postmarked 1906 and was sent to James Gordon who was at that time the Brewer at Cardow distillery, today more familiarly known as Cardhu, the home of Johnnie Walker.

According to unpublished research undertaken by Professor Michael Moss of Glasgow University the partners planned that their distillery would cater largely for the bottle trade in 'self' or single malt whisky, perhaps harking back to Moir's roots as a general merchant, but needed to find a market for the large proportion of surplus capacity.

They were fortunate to find a customer in the prestigious Glasgow firm of wine and spirit wholesalers, Robertson & Baxter (R&B), who will later feature dramatically in this story. R&B already acted as agents for the Greenock Distillery Co.; McMurchy, Ralston & Co., owners of the Burnside distillery in Campbeltown (not to be confused with the earlier Portsoy distillery of the same name) and the proprietors of Fettercairn in the Howe of Mearns in Angus.

They quickly became *Glenglassaugh*'s biggest

customer by a wide margin and in turn supplied the whisky in large parcels to another distinguished Glasgow firm of wine and spirit merchants, William Teacher & Sons. It is pleasing to speculate that *Glenglassaugh* may have ended up in Teacher's celebrated dram shops, drunk with due solemnity!

Though the independent whisky broker has now largely disappeared from the trade, these were bumper times for whisky merchants like R&B who dealt in good quality malt whisky. This was much in demand for the newly fashionable blends that were filling the gap left by the sudden disappearance of brandy after the outbreak of phylloxera and the refusal of the Irish industry to countenance blending, which they considered tantamount to adulteration and fraud.

Little is known about *Glenglassaugh*'s output or other clients in its first years of operation. But, like their competitors in the north-east, it is believed that the partners built up a trade in England where there was a growing markets for 'self' whiskies and, unlike much of the whisky consumed within a year of distillation in Scotland, well matured in sherry wood. Customers south of the border included Wyld & Co. of Bristol, the Birkenhead Brewery Co., Mackie & Gladstone of Liverpool, John Groves & Son of Weymouth, Ernest Hobbs of Gosport, and Richardson Bros of Salisbury.

Glenglassaugh whisky evidently found acceptance in the public houses of Liverpool and surrounds and was equally favoured in the West Country. Most of the firms mentioned were brewers and would have supplied *Glenglassaugh* to their managed and tied estate as the 'house' whisky.

Within three and half years of opening, on 28th December 1878 the distillery was the site of a tragic accident. The stillman, George Aitken and his colleague John Grant, one of the distillery labourers, chose for some reason to clean one of the washbacks immediately it had been emptied. To do this, they would have climbed into the vessel carrying heather besoms, with which to scrub the wooden sides clean of yeast residues.

Having worked at the distillery since it opened, Aitken would have known that the washback contained dangerous levels of carbon dioxide gas, a by-product of fermentation, which is of course fatal. Contemporary reports of the incident express mystification at the men's failure to test the air in the deep washback using a lamp or other open flame, as was standard practice. My own speculation is that they were anxious to start work before the sides of the vessel dried and were harder to clean, though there is no evidence to this or any other effect.

Both were swiftly overcome by the fumes but their cries were heard by the brewer Fraser. He bravely went to their assistance but could not lift either man clear so managed to scramble out and seek further help. Having done so, valour overcame discretion and he returned to the washback where he himself promptly collapsed.

There were now three insensible men at the bottom of the washback and quite a crowd gathered next to it, including Mr Allan, Manager; Mr Tolmie, officer of the Inland Revenue; Mr Grant, clerk at the distillery and others.

According to the *Banffshire Journal,* '*the gas was felt exceedingly strong at the mouth of the tun, so much so that a lamp would not burn*'. Notwithstanding this alarming signal, Adam and Alexander Ingram, described as maltsmen, volunteered to go down with a rope attached to their persons to assist their colleagues. First the one and then the other tried, but they no sooner reached the bottom than they had to be pulled up. Twice each tried with the same result.

These rescue attempts having failed it was decided to cut a hole in the base of the washback. The staves were some three inches thick however and, despite energetic efforts with a saw, axe and hammer, this took some 20 minutes, after which the unfortunate men were rapidly pulled through a two foot square gap— Fraser first, Grant next and Aitken last. They emerged *'…much exhausted, but alive. George Aitken, we regret to say, was dead. Drs Anderson and Robb were at the Distillery a few minutes after the men were taken out. They could do nothing for Aitken, but through their care the other two men have recovered.'*

Aitken, 47 years of age, left a widow and seven children. The incident was said to have cast considerable gloom over the district. Just as their contemporaries, we can regard this tragic accident with some bewilderment: the men had no business entering the vessel without first checking the CO_2 levels and Aitken in particular should have been alert to the dangers. His alacrity to start work on cleaning the washback came at a high cost.

The incident also tells us something of nineteenth century distilling: like much manual labour, it was arduous and dirty, much of the work was repetitive and occasionally dangerous. It is easy to imagine some 'golden age' of artisanal distilling and romanticise a nostalgic view of a lost age of innocence. Aeneas Macdonald, for example, writes of the *'relics of a vanished age of gold when the vintages of the north had their students and lovers'*. Few of us, however, would be happy to swap places with the like of George Aitken for more than a few hours of experience tourism before returning to twenty-first century comforts (and drams) with some relief.

We may today look askance at distilleries run by computers, though *Glenglassaugh* is doggedly traditional in generally eschewing such aids, but understandably rigid health and safety procedures require a risk assessment of such an inherently hazardous procedure, rendered unnecessary in any event by in-plant cleaning. Today's George Aitken would have obtained a 'permit to work' from his distillery manager or supervisor. A gas analyser would be used to check the CO_2 levels and a washback, or any such vessel, would only be entered after stringent checks and with great circumspection. Alert to the dangers, today's employees take considerable care not to drop any personal belongings into washbacks—distillery visitors should be similarly cautious!

The Census records of 1881 reveal the population at the distillery. Dugald Mathieson, the Manager, lived with his wife, four children and one servant in a stone-built modern villa provided by the company. Ten years later Mathieson was still in residence with—by today's standards—the remarkable total of nine further persons under his roof. But shortly after that Dugald Mathieson died and his place as Manager was taken by Alexander Christie who had started with the company as a clerk.

The 1823 Excise Act, which did so much to shape the modern industry, required the distiller to provide a house for the resident Excise Officer, or 'Gauger' as he was known. Though the Government paid a rent not exceeding £10 per annum, nothing was said about repairs and the Gauger's house was not always in tip-top condition. However, the Excise Officers' accommodation at *Glenglassaugh* appears to have been of a higher standard, Barnard describing them as *'handsome dwelling houses, with large gardens'*. Today they are used as the company's offices.

In 1881, they housed James Wilson, his wife and four children and next door his colleague Murdo Tolmie, his wife and young son. Also on site was Robert Sommers, the Brewer (Assistant Manager) with a further five persons in his household and up the road at the Black Jug we would have found William Henderson, a maltman, with a family of four. Finally, a property known as Claylands was home to Alexander Ingram (the would-be rescuer of the unfortunate

Aitken), another maltman. A total of nine persons occupied the accommodation here.

Further members of the workforce presumably lived further afield and were not included in the census for *Glenglassaugh* but the figures for 1881 show a total of forty adults and children in the immediate vicinity of the distillery. No doubt the small glen and the Fordyce Burn rang to the cries of the children and *Glenglassaugh* must have been a very lively place indeed.

Thomas Wilson the Portsoy still manufacturer and junior partner in the venture, died in 1883, followed four years later in 1887 by Colonel James Moir. With no children, he left his substantial fortune of £23,000 (close to £2m today) to his niece Margaret, wife of Alexander Morrison. A statement of his affairs showed that he had an investment of £5,128 (c£424,000) in the distillery and whisky stocks of £862. Clearly, *Glenglassaugh* had been a spectacular success, the distillery shares and whisky stocks representing more than a quarter of his ample estate.

During the last six years of his life he had been unstinting in his public services, chairing the committee to restore the parish church in 1881, leading the campaign to rebuild the harbour in 1885 and supporting the building of a new Town Hall in 1887.

The whisky industry had encountered some temporary difficulties in the mid-1880's but, emerging from these setbacks and the disruption following Moir's death, Alexander Morrison decided to re-equip the distillery. The original plant had run successfully for some twelve years and presumably was ready to be overhauled in expectation of an even more prosperous future. Accordingly, he began the process of expanding and renewing the distillery.

During 1887 the two stills were replaced and, in the same year, Alfred Barnard (see Chapter 6) made his first visit to *Glenglassaugh* on behalf of *Harper's Weekly Gazette* for his monumental work

The Whisky Distilleries of the United Kingdom. Judging by his account, the replacement of the stills had taken place by the time of his visit. He describes the still house as a *'modern edifice'*; the wash still having a capacity of 4,000 gallons (18,160 litres) and the spirit still 2,000 gallons (9,080 litres). This compares to the earlier report of a 2,500 gallon wash still and a 1,500 gallon spirit still.

They were, of course, direct fired. The manufacturer of the new stills has not been recorded, nor is it clear if the old pattern was copied. What is evident, however, is that the distillery was trading sufficiently well to justify a significant expansion.

Barnard records *Glenglassaugh* as *'steadily gaining favour in the market'* with an annual output of 80,000 gallons (363,200 litres). By contrast, the same author

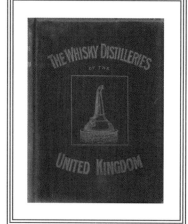

relates that Edradour in Perthshire was then producing 6,600 gallons annually; Glenmorangie 20,000 gallons; Laphroaig 23,000 gallons and The Macallan (which receives a scant eight lines, as opposed to the two pages describing *Glenglassaugh*) about 40,000 gallons. By contrast, however, the grain distilleries at Bo'ness and Port Dundas in Glasgow were producing 870,000 and 2,562,000 gallons respectively, dramatically illustrating the impact of blending on the industry.

Barnard does not appear unduly impressed by *Glenglassaugh*, though he does not dismiss it particularly briefly either. A full two pages of description is relatively generous but, sadly, there is no illustration in his *magnum opus*. However, he was to return some eleven years later and wrote a more detailed account, which is fully discussed in chapters 2 and 6, following.

In 1889, Morrison continued his improvements and two new washbacks were installed. Three years later, in the summer of 1892, the existing three washbacks were rebuilt and elevators installed to raise malt to the kiln, and two barley separators acquired for screening and grading the barley. With the whisky boom gathering volume and demand for whisky strengthening, the outlook appeared excellent. *Glenglassaugh*'s future prosperity, and that of its fortunate owners, must have seemed assured.

Quite unexpectedly, however, in the late summer of 1892 William Morrison died and in September 1892 Alexander Morrison, needing to find the cash to settle his estate, decided 'for family reasons' to sell the business. He wrote at once to Robertson & Baxter to enquire if they were interested.

Glenglassaugh's history would take a dramatic turn.

CHAPTER 2

Enter 'The Highland'

What was he thinking of? I am entirely mystified by Alexander Morrison's haste in disposing of the distillery. It had recently been refitted; sales were good; the whisky market was booming and the family was well off. Even if he personally lacked the necessary resources, with the distillery and stocks of whisky as security Morrison would have had little difficulty in raising a loan to meet the expenses that apparently so concerned him; alternatively, it would at this time have been a relatively straightforward matter to find new investors in an apparently flourishing going concern.

The years 1891–93 were years of plenty in Scotch whisky, with distillery construction going on apace. In that short period Craigellachie, Strathmill, Balvenie, Glen Mhor and Knockdhu were all opened—all of them classified as Highland distilleries. Further hectic construction followed until the abrupt ending of the boom with the failure of Pattison, Elder & Co. of Leith in December 1898.

Morrison cannot have been unaware of the fortunes being made in Scotch whisky in the 1890s. However, despite this and whatever his motives he moved immediately to sell *Glenglassaugh*.

Even allowing that he had his reasons he handled

the business badly. Rather than advertising the sale and exposing *Glenglassaugh* to the market he simply wrote to his main customer— hardly commercially wise—to see if they were interested.

Robertson & Baxter wasted little time. The two principal partners William A. Robertson and J.C.R. Marshall travelled north from their spacious offices at 48 West Nile Street, Glasgow to discuss the matter with Alexander Morrison's lawyers, Paul & Williamson of Aberdeen, on 13th September 1892. There they promptly expressed interest in buying *Glenglassaugh* for their sister company, Highland Distilleries, which had been incorporated in 1887 and shared both several directors and office premises with R&B.

The meeting evidently went well and, as a result of their conversation, Alexander Morrison replied the next day setting an asking price of £15,000, with a commitment to continue buying all his supplies of whisky from the distillery. This makes his haste all the more perplexing—had Morrison intended to leave the liquor trade entirely out of religious or other conviction, or had he wanted cash to pursue a life of luxury and dissipation in Monte Carlo or some other resort, his alacrity could be understood, if not condoned, but for someone intending to continue in the whisky business, at least in part, he seems to have proceeded with an extraordinary and inexplicable combination of naivety and speed.

Robertson and Marshall were evidently shrewd negotiators; more than equal to Morrison and his advisers and, no doubt, alerted to a bargain by the enthusiasm of his response to his willingness to conclude the business. However, rather than immediately accepting the offer they requested a technical survey and a week later William Grant, the manager at Bunnahabhain on Islay, was despatched to Portsoy to make a thorough technical survey of the distillery.

Bunnahabhain had been built by William Robertson and J.C.R. Marshall in 1881 and was glowingly described by Alfred Barnard as containing plant *'of the newest and most approved description'* in a building both compact and systematic. At this time it was producing around 200,000 gallons of whisky annually, there being considerable demand for Islay make in blending. William Grant, therefore, was likely to be a sound judge of both the distillery layout and the quality of the spirit.

In fact, R&B and Highland were discriminating, not to say demanding, in their requirements. Between 1887 and 1891 they had been offered and considered purchasing Ben Wyvis, Bowmore and Fettercairn distilleries: none met their very exacting standards. Their sustained interest in *Glenglassaugh* is therefore suggestive of both its quality and potential—factors of which Alexander Morrison appeared blithely unaware and apparently did not trouble himself to determine. In today's terminology, this was a particularly slipshod piece of due diligence on his part—negligence which was to prove an expensive error.

Grant's report was enthusiastic and he concluded *'After due deliberation and considering (the) convenience and stability of the whole works if the produce is of good marketable quality to the best of my knowledge and belief the place is worth from £14,000 to £15,000'*.

Here again, however, water enters the *Glenglassaugh* story, and once more to significant effect. The cautious and prudent William Robertson asked for samples of the mashing, reducing and cooling water used at *Glenglassaugh* to be sent to Glasgow for analysis.

The results of these tests were far from satisfactory: one sample was discovered to contain oxidised sewage products and another was considered to be *'much too hard for either mashing or reducing'*. Only one of the samples was thought to be suitable. As a result Robertson & Baxter were only prepared to offer £10,000 in cash for the plant, excluding the stocks, which Alexander Morrison grudgingly accepted. Quite why he did not have the water samples tested himself, or dispute the findings, or attempt to negotiate the price remains a further mystery in what is a frankly puzzling episode in *Glenglassaugh*'s history.

With the deal completed Robertson & Baxter promptly capitalised on their good fortune: coolly inviting Highland Distilleries to buy *Glenglassaugh* for £15,000. In other words, in short order they were looking to turn a 50% profit of £5,000 on the deal—or around £400,000 in today's terms. A nice piece of business if you can get it, one might well conclude.

Quite cognisant of what was going on, the two local Highland directors Robert Dick and William Grant, who were based at Glenrothes distillery, hurried off to examine the plant. They were delighted with the place, if alert and perhaps a little sensitive to the hard deal struck with Morrison.

'The deeds should be prepared with as little delay as possible and as we know that R&B are making a profit on this action, which they well deserve, the actual price paid should be the consideration price stated in the deed of conveyance in case the Morrisons might kick did they see that the HD Co. were paying R&B more than they purchased the distillery at. I should not mind if they (R&B) made £5,000 out of it. It is cheap to the HD Co.

It is substantially built and replete with everything, with ample warehouse accommodation for years to came'.

Was this whole affair sharp practice or simply shrewd negotiation by two determined and hard-nosed men of affairs? We may imagine Mr Morrison's eventual reaction!

The Highland board agreed to buy in November 1892 and during the following month *Glenglassaugh* passed under the company's control. With improvements designed to increase capacity and at the same time reduce costs, output climbed to over 110,000 gallons in the boom year of 1898. Thereafter production dwindled to less than 28,000 gallons in 1907.

Some records from this period survive in an archive at Glasgow University and several fascinating facts emerge. Highland were clearly satisfied with the management of the distillery. A young man, scarcely in his twenties, Alexander Christie had taken over this position from Dugald Mathieson on Mathieson's death. He was kept on in this role until he himself died in September 1902 at the age of 32 following a fall from his horse and carriage.

Christie's premature death was the good fortune of Dugald Mathieson's son Edward: references in September 1892 from Alexander and James Morrison and another from his father from February 1893 suggest that he was seeking to leave *Glenglassaugh* to better himself (*'to our regret'* wrote the Morrisons) and thus the unlooked-for vacancy represented a timely opportunity of promotion.

But family history records that in an idiosyncratic move for one aspiring to a management position in distilling, Edward Mathieson took a pledge of total abstinence from the wonderfully-named London

Road United Presbyterian Church & Mission Total Abstinence Society in June 1896. This required him to eschew *'all intoxicating drinks as beverages'*. One can only speculate on how spirit quality in the distillery was maintained but perhaps his conscience permitted discreet tasting for professional purposes, provided he did not enjoy himself.

It appears to have proved little obstacle to a successful career, however. He remained in his post at *Glenglassaugh* until 1907 when he transferred to Tamdhu, serving there for a further twenty years.

The Barley Delivery book for 1893 also survives and contains several features of interest.

Firstly, as might be expected, most of the barley purchased for the 1893/94 distilling season was local: purchases were made from various farmers in Portsoy at prices ranging from 24/- (£1.20) to 25/6 (£1.27½) per quarter. Barley was also sourced in Inverness, Aberdeen and elsewhere. Two curiosities stand out, however.

Around one-third of the barley purchased, at least on the pages surviving from the ledger, is described as 'Russian barley'. Clearly the quantity available played a part, but also price. Despite the costs of transport, *Glenglassaugh* was able to buy Russian grain at 22/3 (£1.11) a quarter, a substantial saving on the cost of Scottish barley. If nothing else, this is an interesting insight into distillers' purchasing habits a hundred or more years ago at a time when the complaints of Scottish farmers regarding imported barley are regularly in the news!

And there is something else. At the time of the sale the distillery and surrounding farm were separated. The records show that the highest price paid by *Glenglassaugh* was for grain from Craigmills Farm, which commanded 27/- (£1.35) for a quantity of 97⅝ quarters—a considerable premium for grain that was literally just up the road. May one detect just the slightest sense of guilt or conscience in this?

Whatever the truth of that, it's clear that the Highland Distilleries were proud of their stable of distilleries, so much so that in 1898 they commissioned the renowned Alfred Barnard to visit their four distilleries and write a promotional pamphlet extolling their virtues (see Chapter 6 for further details and a facsimile of the section on *Glenglassaugh*). Barnard was, by then, quite an authority in the drinks trade having written not only the magisterial *The Whisky Distilleries of the United Kingdom* but the even more impressive four volumes of the *Noted Breweries of Great Britain & Ireland*.

He was happy to accept commissioned work, having recently completed another money-spinning assignment for William Whiteley the London retail entrepreneur and founder of Whiteley's department store in Bayswater. *Orchards and Gardens Ancient and Modern* (1895) was a commercial disaster for Whiteley but no doubt lucrative for the author, by now ensconced in Glenalmond, his handsome villa in South Norwood.

He accepted the chance to revisit Scotland and his beloved distilleries with some eagerness, commenting:

'*During the past ten years* (i.e. since the appearance of his *Distilleries*) *the manufacture of whisky has increased by leaps and bounds until it has become a gigantic industry. Scotch whisky is daily penetrating regions hitherto unknown to commerce... the leading beverage and source of profit to the enterprising merchant who dispenses it.*'

The Still House, c. 1898

The pamphlet is now rare and with its extensive photographs of the still room and buildings, an important reference source for the appearance and layout of the original distillery. Accordingly, it is reproduced in facsimile following Chapter 6.

Records survive for production and yields from 1902 to 1907. Over the period, both output and efficiency rose steadily from around 210 litres of absolute alcohol per ton of malt in 1902 to 286 laa/ton in 1907. Production fell off dramatically however in 1907 and the distillery was closed that summer, though malting continued on the site until 1922.

The opening years of the twentieth century were difficult ones for Scotch whisky. The failure of Pattison, Elder & Co. in 1898 with the attendant revelations of blatant fraud in their blending brought down ten other firms and led to a slump in prices. Though Highland Distilleries had done very little business with Pattison's (Walter Pattison's flamboyant personal style being anathema to the more circumspect William Robertson), both they and Robertson & Baxter were affected by the general downturn. R&B closed their Leith offices and the combined Glasgow office of the two firms moved to 106 West Nile Street.

This was compounded by a general deterioration in the UK economy following the end of the Boer War. Output had been running ahead of demand for some years in the late 1890s and though Robertson & Baxter secured good order for fillings in 1902/03 and for the following two seasons they could not avoid the general malaise.

The 'What is Whisky?' case and subsequent Royal Commission (1908) led to great uncertainty in the trade and this loss of confidence, combined with the drop in demand, contributed to *Glenglassaugh*'s demise. The distillery was, in the end, a casualty of wider economic conditions.

Malting continued at *Glenglassaugh* until 1922,

presumably for Highland's Glenrothes and Tamdhu distilleries, when that too was suspended. Duty increases in the 1919 and 1920 Budgets had dramatically reduced the distillers' profits, the Government preventing the trade from passing the full increases onto the consumer. Consumption fell steadily after 1920 and with it, demand for malt.

Glenglassaugh was then completely silent, and a shadow of its former self. It must have presented a depressing prospect and a sad contrast with the bustle of the 1890s. Though some of the buildings were let out for storage—Hutcheon of Turriff, agricultural merchants, using the warehouses for example—activity was at a low ebb. Accordingly, there is little to record in the years between 1907 and 1960, apart from one curiosity.

In his generally authoritative *Scotch Whisky Industry Record* Charles Craig has it that *Glenglassaugh* operated in the 1931-32 season and then again from

WHAT IS WHISKY?

ASK A POLICEMAN.

1933-36. However, in his unpublished account of Highland Distilleries, Professor Michael Moss describes the distillery as silent from 1907 to 1960, when it was entirely remodelled.

It would be quite remarkable if *Glenglassaugh* had worked at this time, for these were the darkest days of the Scotch whisky industry, and indeed the country as a whole. In 1933 whisky production dropped to its lowest level for more than 100 years and just fifteen distilleries were working in the whole of Scotland. This was a bleak and depressing period.

Both Craig and Moss would normally be regarded as impeccable sources and it is curious to find them at variance on this historical detail. If *Glenglassaugh* did distil in the 1930s it would be a matter of some interest but, on balance and being unable to trace any documentary evidence, I am inclined to Moss' view. Further research is certainly needed, but it would indeed be a pleasant surprise to find that this little distillery, neglected and forlorn since 1907, had enjoyed a brief renaissance during the Great Depression.

John Mitchell and his mother, Distillery Cottages, 1940

Whatever the truth, *Glenglassaugh* underwent further enforced change during the Second World War. Sandend Bay was considered a potential landing spot for an invasion force, so substantial concrete blockades and gun emplacements were erected to cover the beach and defend the track through the distillery. The distillery itself was taken over by the military and various buildings pressed into wartime service.

Most were occupied by the NAAFI, who established

a bakery in the easternmost maltings building and also erected a warren of small offices in both maltings, thus preventing further use of the malting floor. During clearance work in 2008/09 a considerable amount of wartime graffiti was discovered, some of which has been carefully preserved as part of the building's history. Apparently, active service at Sandend Bay was not an unduly onerous, demanding or dangerous posting—or even a particularly active one, though a constant 24-hour vigil was maintained on the beach for any sign of German landing craft, attracted perhaps by the smell of early morning rolls floating across the beach.

Parts of the site were also used by the Signal Corps and the Home Guard and a plan of their works survives at the distillery today. At that time, the 1875 distillery and the lower warehouses were still standing and there was room for all the various occupants to come and go as they pleased. But, after 1945 and the withdrawal of the troops, *Glenglassaugh* went back to sleep.

Little of note then occurred for the next fifteen years. Encouragingly, Highland Distilleries re-opened Tamdhu in 1947 following twenty years of closure, though it had briefly distilled during the 1941/42 season.

Whisky output began to grow from 1950 and the pace of expansion and recovery accelerated towards the end of the decade as confidence grew with economic growth and strengthened exports. A considerable number of distilleries were re-opened, expanded or built around this time, including Strathmore (Cambus), Invergordon, Tormore, Isle of Jura, Glenfarclas and others. In fact, nothing quite like this expansion had been seen since the 1890s and *Glenglassaugh* was to be a beneficiary of this optimism.

The mood was one of confidence and a belief in the progress of science. Accordingly, when Highland Distilleries' Engineering Department came to look at *Glenglassaugh* it must have seemed an easy, indeed comfortably radical decision, to demolish the old distillery and build a completely new one. The old was to be cast aside and replaced by everything à la mode. This was the Sixties after all, if only just, and the times they were a-changin'—even in Portsoy.

So everything beyond the northern boundary of the malting, save a solitary cottage, was razed to the ground. Mill and mash room, distillery, spirit store and the lower warehouses were all swept away and the ground cleared.

Moving barrels, c.1970

Evidently, the thinking was that the distillery was small, obsolete and old-fashioned. Indeed, it may well have been impractical to restore to working order after more than 50 years in mothballs, but aerial photographs from the period seem to show the warehouses in good order. Perhaps that impression is misleading, or perhaps it was considered awkward and inconvenient to transport casks the short distance down the track to the old buildings. In any event, between 1960 and 1962, they too came under the wreckers' unforgiving ball.

Planning to expand production and build stocks for blending, in 1959 Highland built a substantial new racked warehouse to the south of the distillery, between the original dunnage warehouse (today known, appropriately enough, as the 'Number 1') and the main road, the A98. Though the dunnage warehouse was still in use, additional racked capacity was soon required and further modern buildings were erected in 1961 and 1963. The site now can hold almost 30,000 casks in the racked warehouses with additional capacity in Number 1. Though this suffered a partial roof collapse in 2007, it is now being brought back into use and houses a variety of cask types, including some experiments with barrels formerly used in various wineries.

A small filling store, cooperage and workshop were also built. Spirit was pumped via an overhead pipe to the filling store and casks transferred a few yards to the warehouses. It is certainly a great deal more convenient than would have been the case had the old warehouses towards the shore been retained. The filling store has since been completely refurbished (the party to celebrate the distillery's re-opening was held there) and is in use today, though there are currently no plans to re-instate the cooperage.

The Filling & Cooperage Squad, c.1970

With an aim of retaining virtually all *Glenglassaugh*'s production for sale as single malt, and thus harking back to Moir's original vision, Nickerson and Eunson have taken a doggedly traditional approach and prefer to use the old dunnage space. Out of common prudence, however, some spirit is also being stored in the racked warehouses and, in time, will provide a fascinating opportunity to compare the impact of alternative methods of storage on the resulting spirit character.

The building of the new racked warehouses and the move of the still house to a position of some prominence overlooking the bay rather changed the character of the site. Today the entrance is dominated by warehousing while the older buildings, further into the site, come as something of a contrast and a welcome surprise to traditionalists especially after the somewhat stark 1960 office that confronts you on arrival.

The new distillery was linked to the western of the two surviving Victorian buildings which housed the malt intake, malt bins, mill and brewer's office (now disused). This was connected by a tunnel carrying the grist by means of an Archimedes' screw to a dramatic new building with a striking serrated roof, intended presumably to echo the line of waves rolling onto Sandend Bay. This could not have been built at any time other than the 1960s and is as bold as it is assertive. Craig describes the new distillery as *'a very advanced design for its time'* and this refers equally to the building and to the still room.

For all that, the new distillery was on a relatively modest scale with just one pair of stills, capable of producing around 220,000 gallons (1m litres) of spirit annually, or slightly more if worked hard. The design of the stills and the look of the still house are, however, radical for the period. Vertical shell and tube condensers were used, instead of traditional worm tubs and this, together with the overall appearance of the still house is strongly reminiscent of the work of William Delmé Evans, the Welsh distillery designer and engineer.

I am not suggesting that he designed *Glenglassaugh* but his innovative theories and designs were highly regarded in the industry at the time, following his work at Tullibardine and the Isle of Jura distillery. It is a little known fact that, around 1960, he was working for the blenders Brodie Hepburn on the construction of the Macduff distillery (now operated by John Dewar

& Sons) not ten miles from Portsoy. He never referred subsequently to this work, having left the project prior to its completion following an unspecified disagreement with his clients but photographs of the Macduff still house prior to its reconstruction in 1968 bear a striking similarity to *Glenglassaugh*, as does the Tullibardine design to this day.

Later Delmé Evans designed Glenallachie where the angular roof bears more than a passing resemblance to the saw tooth detailing at *Glenglassaugh*. However, the formal credit belongs with the Engineer's Department at Highland Distilleries head office in Glasgow.

Not everything was new, however. The Porteus mill and malt handling equipment was second-hand. The mill, in fact, was dated as pre-World War II by the engineer who refurbished it in 2008 and the unusual cast-iron Porteus mash tun is also of an earlier vintage. Both mill and mash tun appear to have years of life ahead of them, a tribute to the quality of Porteus' engineering and construction. The original six wash backs were constructed in Corten steel, though later these proved unsatisfactory.

Apart from the still house there are some other unusual features at *Glenglassaugh*, most notably the layout and the use of an underback below the mash tun. The layout is something of a surprise in a brand new building: everything proceeds logically and in a more or less straight line from the malt intake to the mill to the mash tun, after which one would expect to find the washbacks and finally the still room. However, the latter two components are in reverse order, meaning that green wort is pumped past the still room and, after fermentation, wash pumped back to it.

This has the aesthetic advantage of placing the still room towards the centre of the building, which is architecturally more pleasing and allows for a somewhat more dramatic entrance to the main distillery building. Though this layout seemed to me both illogical and confusing the present generation of distillers were happy to enlighten me. Though it requires a little more pumping of liquid the distillery is evidently much easier to operate with the key functions all on one level. In fact, Stuart Nickerson says he would have designed it that way had he been in charge!

One agreeable facility, however, is a high-level link from the main floor of the washbacks to a gallery above the still room. Visitors can gaze into the still room to their heart's content, take pictures and study the process in complete safety, without intruding on day to day operations, though tours normally continue into the still room to inspect the original spirit safe, handsomely restored by Forsyth's of Rothes.

The use of an underback on the mash tun is surprising in such a consciously forward-looking design but it has now been abandoned in favour of a balanced drain system. The underback remains in place simply as a historical curiosity. Taking tours round the distillery, however, Stuart Nickerson cannot resist pointing it out and recalling that he was personally responsible for painting it only hours before Alex Salmond arrived to re-open the distillery. Perhaps for that reason alone its survival seems assured, even if it is now functionally redundant.

The first Manager at the brand new *Glenglassaugh* was Pat Smith who was provided with a new house on the site. This stands to the west of the main distilling complex and is, frankly, as ugly and impractical as the distillery is modernist and striking. Apparently, Smith had a hand in the design but his successors were reputedly less than charmed by the shortcomings of their accommodation.

Presumably the design is intended to echo that of the new distillery. It is, however, an architectural disaster. The radical design does not work on a domestic scale and little consideration appears to have been given to the orientation of the property on the site. With the opportunity of spectacular views looking from a prominent position over Sandend Bay to Sandend village itself, the architect elected to construct a blank curtain wall where the location cries out for panoramic picture windows. Perhaps the builder put it the wrong way round!

Today it is used as a dry goods store and small hand bottling line, one of the current operation's proud claims to fame being that all *Glenglassaugh* whisky is bottled at the distillery.

Blended whisky was very much in the ascendancy in the 1960s and 1970s. Robertson & Baxter were major suppliers to the trade and also had their own blends through their subsidiary, Lang Brothers which they acquired in 1965. Together with Langs Supreme, Red Hackle was a particular favourite and, more significantly, R&B had an interest in Cutty Sark, a leading brand in the U.S.A. They had blended this for London wine merchants Berry Bros & Rudd, since

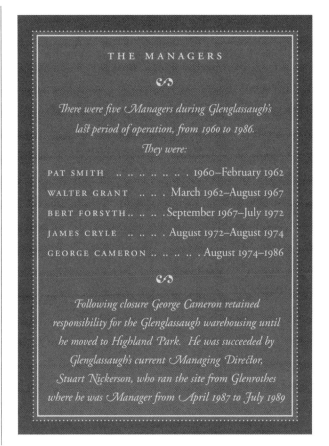

THE MANAGERS

There were five Managers during Glenglassaugh's last period of operation, from 1960 to 1986. They were:

PAT SMITH 1960–February 1962
WALTER GRANT . . . . March 1962–August 1967
BERT FORSYTH . . . . . September 1967–July 1972
JAMES CRYLE . . . . . August 1972–August 1974
GEORGE CAMERON . . . . . August 1974–1986

Following closure George Cameron retained responsibility for the Glenglassaugh warehousing until he moved to Highland Park. He was succeeded by Glenglassaugh's current Managing Director, Stuart Nickerson, who ran the site from Glenrothes where he was Manager from April 1987 to July 1989

1936 and eventually sourced around 70% of the single malts in Cutty Sark from Highland.

In 1970 Highland acquired Matthew Gloag & Son Ltd of Perth, owner of The Famous Grouse brand and entered into a joint venture agreement with R&B with the object of developing The Famous Grouse into a major international brand. In this they were strikingly successful, sales of Grouse, as it is colloquially known in Scotland, rising from 40,000 cases at the time of the takeover to more than 1,000,000 cases by 1979, when Highland was the subject of an unsuccessful 'merger' approach from Hiram Walker of Canada, owner of the Ballantine's brand. Grouse has, of course, continued to grow strongly since then.

So throughout this period, the focus of *Glenglassaugh*'s owners was very much on blending and, from 1960, virtually all of the output was required for the ever-increasing Grouse and Cutty Sark blends. Fillings were also supplied for J&B Rare and some small brokers placed modest filling contracts. But there was a problem.

Not only was *Glenglassaugh* significantly smaller than Bunnahabhain, Glenrothes, Tamdhu and Highland Park (which had been acquired in 1937 as an integral part of the Cutty Sark blend) but its harder water made spirit in a distinctively Highland style.

Water again!

Under the direction of R&B's forceful John Macphail, the blenders at R&B Jack Maclean and Paul Rickards, favoured a softer Speyside style. This was certainly the style looked for in Cutty Sark, with its pale colour and light flavour. Accordingly, considerable efforts were made to get *Glenglassaugh* to produce spirit of this type. Distorted, second-hand recollections of these attempts have led to the one great myth about *Glenglassaugh*: that there is a 'problem' with the water, a canard that has been carelessly repeated in more than one whisky guide book.

So let us set the record straight, straight from the mouth of Jim Cryle, then Manager (1972-74) at *Glenglassaugh*. Cryle came to *Glenglassaugh* as a relatively young and ambitious man. This was his first managerial position, having previously been Assistant Manager at Bunnahabhain on Islay. Now retired after a distinguished career, he recalls that *'this was very much a trial posting for a young manager'*. After leaving Highland Distillers in 1991 he went on to join Chivas Brothers at The Glenlivet, perhaps the most famous of all Scotch whisky distillers and was long renowned as their Master Distiller.

He arrived at a challenging time. The R&B blenders were not happy with the spirit from *Glenglassaugh* and preferred the Glenrothes style. Accordingly, with the assistance of Dr J.D. Gray of Inveresk Research International and later Pentlands Scotch Whisky Research Ltd (in which R&B had an interest) various experiments were carried out.

These involved bringing water for mashing, fermentation and distilling from Glenrothes to *Glenglassaugh* by tanker to assess the impact of this on spirit quality. Then *Glenglassaugh* water was used to mash at *Glenglassaugh* and the fermented wash tankered to Glenrothes to be distilled and the spirit character compared. Glenrothes water was used for mashing and fermentation at *Glenglassaugh* and the wash transferred back to Glenrothes for distillation—and so on. A number of variations of this experiment were undertaken, all involving tankers of various liquids rolling on and off the *Glenglassaugh* site—little wonder that the idea grew up amongst the curious but uninformed that there was a 'problem' with *Glenglassaugh*'s water. Changes were also made to the cut points—the point at which the stillman selects the middle cut of the spirit that will eventually be filled to cask. Meticulous records were kept of the experiments and these still survive. It all sounds very laborious, but thorough and painstaking.

The result of all this was that it was determined that the best fit to the blenders' requirements was to replace the stills at *Glenglassaugh* with a pair styled on the Glenrothes stills, but to continue the use of local water from the Fordyce Burn, which continued to flow as reliably and as cleanly as ever.

Jim Cryle remembers the old stills as having *'more of a tulip-bulb, straight-sided shape; no waist to speak of and no boiling balls'*. Previously, in another experiment to influence the spirit quality, charcoal filters had been fitted off the lyne arm of the spirit still, the blenders having declared the spirit 'too oily' for their requirements. These were not replaced in the new design.

Forsyth's of Rothes supplied the new equipment in February 1974, which required the roof to be removed before the old stills could be removed and new ones installed. It is, in fact, a challenge of the building design which rather looks as if it was built around the plant, with little thought given to the question of eventual replacement.

The new stills did not represent any increase in capacity for the distillery, this being constrained by the size of the mash tun and available fermentation capacity. However, Jim Cryle recalls that, at its busiest, *Glenglassaugh* was filling close to 1.5m litres of alcohol a year—which must have been hectic! However, with a total of sixteen staff on site and a floating shift system, the spirit was produced and safely stored away in the giant warehouses.

The Corten steel washbacks were also proving problematic, the inside surface of the soft steel having become pitted and thus difficult to clean. Eventually, two of the vessels were replaced with wood but in calculating the capacity, the surveyors forgot to allow for the thickness of the wooden staves compared to the steel and the new washbacks proved too small.

Extension pieces were fitted, which explains the curious height of the washbacks off the floor of the room. Later, when two further wooden washbacks were installed, the height was matched. The two remaining Corten steel washbacks were lined with stainless steel. Eventually they were replaced by stainless vessels which remain in place to this day, though not currently in use.

The company clearly regarded Cryle as having successfully managed this trying time at *Glenglassaugh* and when, in July 1974 a vacancy arose at Tamdhu he was promoted to the position of General Manager of that site, running both the significantly larger distillery and the associated maltings. George Cameron, known to all as 'Dod', took his place.

Virtually all the spirit was filled then to refill hogsheads and butts, with little or no sherry wood being used—it would not have fitted the Cutty Sark or Famous Grouse style and almost all the fillings were required for blending. However, it is now clear from the award-winning quality of the remaining mature whisky that the wood quality must have been high. At a time when wood management was cavalier in many companies, Jim Cryle remembers the R&B policy very clearly. *'They always insisted on good wood—any hint that it was tired and out it went.'*

In general the mature whisky would have been used at 6 to 8 years of age. *'In those days'*, Cryle told me *'the tendency was to use more mature whisky than today and the R&B blenders were very choosy!'*

In a further attempt to influence the spirit character and imitate Glenrothes, water softening plant was installed after 1974 and production continued apace with Dod Cameron at the helm.

In 1980, however, Highland embarked on a major expansion at Glenrothes, building a new still house and adding four new stills to the six already in place. Though apparently not the intention at the time this was effectively *Glenglassaugh*'s death warrant, as wider problems in the industry became evident.

Whisky sales had been buoyant in the 1970s but then began to stagnate. Perhaps excessively controlled from a production perspective, especially at the dominant Distillers Company Ltd (DCL), the industry continued producing regardless and stocks continued to rise—the 'whisky loch' of legend had arrived. By 1980, just as Glenrothes was expanded, industry stocks were equivalent to more than ten years production. Moves to trim capacity began that year when short-time working was introduced, with lay-offs and distillery closures following as the scale of the problem became evident. By 1986 around a quarter of all Scotch whisky distilleries had gone out of production, with the DCL undertaking particularly draconian restructuring.

As in 1908, *Glenglassaugh* was a victim of forces far beyond its control. The distillery's final production run was in November 1986 with the last cask being filled on 3rd December of that year.

Though malt was still stored there, and the warehouses were in constant use, *Glenglassaugh* slumbered under a regime of benign neglect and appears, incredible as it may seem, to have been largely forgotten. From time to time individual casks were bottled by independent merchants but the distillery itself, like so many other victims of the retrenchment of the mid-1980s, appeared consigned to the history books.

In a curious postscript to this history, Highland Distillers was eventually bought at a total cost of £601m by its erstwhile partners Robertson & Baxter, through their parent vehicle, the Edrington Group. Though well financed and financially prudent, this stretched even R&B's resources and required the assistance of another independent company in the industry. Accordingly, William Grant & Sons (owners of Grant's, Glenfiddich and The Balvenie) came in as a 30% shareholder of a new holding vehicle,

The 1887 Company, which acquired Highland and thus *Glenglassaugh*.

In autumn 2001 the *Scotch Whisky Review* asked Ian Good, then Chairman and Chief Executive of the Edrington Group, about *Glenglassaugh*. He replied in few words. *'Closed in 1986, it is now a warehousing complex. I don't see it distilling again.'*

Then, and for quite some time afterwards, few would have disagreed with him.

Aerial view, c.1962

CHAPTER 3

Clinging on — Rare & Unusual Bottlings

Highland Distillers clearly saw no future for Glenglassaugh as a distillery, or as a brand that they would promote, but that does not mean that they were averse to selling the whisky when the opportunity arose.

Though the priority amongst their malt whiskies was, understandably, given first to The Macallan and then Highland Park, contrary to popular belief, *Glenglassaugh* was bottled and sold as single malt following the distillery's closure. It did not receive any significant marketing support, however, and basically relied on the energy and initiative of an individual importer to create what sales they could.

The most successful market was in France, where the importer Hervé Ferté was able to take several thousand cases a year of both a non-aged and a 12 year old official distillery bottling in the late 1980s. This was also sold in Italy and Germany, where *Glenglassaugh* had a modest following, but was discontinued in 1995/6.

Should you discover some it will be in a standard tall round bottle, carrying a simple light tan label featuring a modest line drawing of a horse and cart carrying casks with an artist's impression of the distillery in the background. Humble it may appear, but if unopened do not break the seal: this is rapidly becoming collectable.

However, the distillery's stocks were being run down mainly by continued use in blending and the vast majority of bottled *Glenglassaugh* that now survives is from independent or 'third party' bottlers. Today, this is found mainly in specialist retailers or online auction sites where it is attracting a growing following. My impression is that these older bottles are now circulating amongst collectors and investors rather than being drunk.

Over the years, the independent bottling industry has had a strange and, at times strained, relationship with the distillers. There can be no doubt that independents such as Gordon & MacPhail of Elgin and the Scotch Malt Whisky Society played a significant role in making single malt whiskies available, better known and appreciated at a time when the major companies simply weren't interested. Indeed it's clear that even relatively recently, for many companies single malt whisky was simply a tiresome distraction from the business of building their blended brands.

Independent bottlers are regarded somewhat schizophrenically by the distilling industry. On the one hand, they are viewed as a form of parasite that lives off the efforts of the distillers, abusing their brand values. Some distillers take a militant stance against third party bottlers, taking legal action against anyone using their trademarks and putting great efforts into making sure that product cannot fall into the hands of the independent trade. When casks of whisky are sold for blending the contract will typically specify that the contents will not be sold as single malt, cannot be resold and, if surplus to the blender's eventual requirements, must be re-offered to the original distiller. Another simple measure is to add a very small quantity of a second single malt to a cask of, say, Glenfiddich and sell this as Wardhead, thus precluding the possibility of this cask ever appearing as an independently bottled Glenfiddich.

So effective are these measures that independently bottled casks of famous brands such as Glenfiddich, The Glenlivet, Glenmorangie and so on are simply never seen on the market any longer. A considerable amount of marketing and legal effort goes into maintaining this state of affairs and, with the growth in single malt sales, more and more brand owners are quietly adopting this position.

On the other hand, some distillers have been prepared to take a more tolerant stance. They have seen third party bottlers as a way of making limited quantities of single malt available to enthusiasts, perhaps informally testing the market before offering an official expression or, more cynically, as a way of disposing of unwanted stocks. Perhaps the third party bottlers offer a way to appease single malt aficionados in their quest for endless variety without the distiller being put to the trouble of releasing an official bottling.

Edrington seem to fall between the two camps, making efforts to reduce the supply of The Macallan and Highland Park, so that eventually independent bottlings would simply dry up, yet releasing *Glenglassaugh* fairly freely and appearing relaxed about the use of the distillery's trademark. Since acquiring *Glenglassaugh* the new owners have tightened up in this area and have made energetic efforts to buy back any casks held by blenders or other third parties.

However, with substantial quantities having been sold into the blending market, casks of *Glenglassaugh* were fairly freely available until the acquisition. Knowing that the distillery was out of production and highly unlikely ever to produce again other blenders would have taken steps to remove it from their blend recipes. Over time, their stock diminished and the few casks which remained became increasingly irrelevant. Individual casks could then be disposed of to third party bottlers.

Just as this was going on, a market for obscure single malts began to develop. From the middle of the 1990s, and to a greater extent from 2000 onwards, single malt enthusiasts began actively to seek out the rarer products of lost distilleries. In some cases, such as Diageo's Rare Malts collection, these were even actively marketed by the original distiller but mainly this fell to independent bottlers.

Inevitably, then, the process of selection was random and capricious, depending on a particular cask becoming available, a suitable bottler being interested and the price being right. There was no absolute standard of quality, and some of the reluctance of the distilling industry to supply independent bottlers sprang from an understandable aversion to seeing expensively cultivated brand names being applied to second rate whisky, as was occasionally the case. Not all casks mature well or to a consistent standard and would thus not pass the distiller's own quality control standards for release as a single, though possibly quite acceptable for blending purposes.

Moreover, the boom in the market attracted new would-be bottlers attracted by the fashionability of single malt whisky and the apparent profits to be made: they were regarded with some suspicion as parvenus by a relatively conservative industry where much business still relies on long-established contacts and personal relationships.

The distillery has made a small collection of all the bottlings they have been able to find (both official and independent), sourcing these from on-line auction sites, dealers and private collectors and some of which are illustrated from page 53. With the reopening of the distillery, all are now becoming collectable and there is some evidence of prices rising. Most were released in very small numbers, being the contents of a single cask: as a bottle is opened and drunk, so the rarity of its bedfellows increases.

One curiosity is that many of these bottlings describe the distillery as a Speyside. This misconception seems to have crept in quite early and also appears in various books and on websites. It's wrong: though close to the boundary of the Speyside region, *Glenglassaugh* is and always has been a proud Highland single malt.

These rare and unusual bottlings, some official, represent all that kept the *Glenglassaugh* name in the public eye—however modestly and accidently—from the 1986 closure to the present day. For that reason alone it's unlikely that the modest display at the distillery will ever be drunk. So far as the proprietors are concerned, it's bottled history.

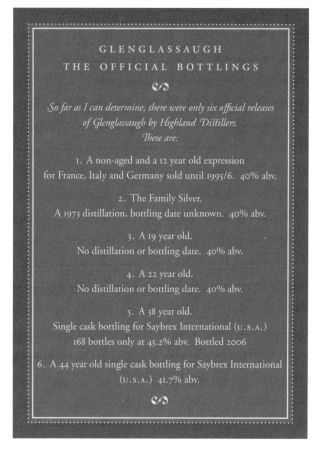

GLENGLASSAUGH
THE OFFICIAL BOTTLINGS

*So far as I can determine, there were only six official releases of Glenglassaugh by Highland Distillers.
These are:*

1. A non-aged and a 12 year old expression
for France, Italy and Germany sold until 1995/6. 40% abv.

2. The Family Silver.
A 1973 distillation, bottling date unknown. 40% abv.

3. A 19 year old.
No distillation or bottling date. 40% abv.

4. A 22 year old.
No distillation or bottling date. 40% abv.

5. A 38 year old.
Single cask bottling for Saybrex International (U.S.A.)
168 bottles only at 45.2% abv. Bottled 2006

6. A 44 year old single cask bottling for Saybrex International
(U.S.A.) 41.7% abv.

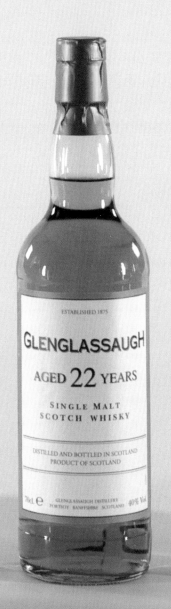

Third-party bottlings so far recorded

Date Bottled	Bottler	Age or Date of Distillation	abv
1984	Gordon & MacPhail	17 yo. 1967	40%
1985	Douglas Laing (D L McGibbons)	10 yo	43%
1991	Cadenhead	13 yo. 1991	59.8%
1995	Gordon & MacPhail	12 yo	40%
1996	Hart Brothers	22 yo. 1974	43%
1996	Scott's Selection	22 yo. 1996	56.9%
1998	Gordon & MacPhail	12 yo	40%
1998	Signatory	31 yo. 1967	55.8%
1998	Scotch Malt Whisky Society	20 yo. 1998	52.0%
2003	The Dormant Distillery Co. (Royal Mile Whiskies)	27 yo	47.4%
2004	The Dormant Distillery Co. (Royal Mile Whiskies)	28 yo. 1976	51.9%
2004	Cadenhead	26 yo. 1978	51.1%
2005	Murray McDavid	40 yo. 1965	47.8%
2005	Dewar Rattray	28 yo. 1976	53.3%
2005	The Highlands & Islands Whisky Company for Jack Wiebers Whisky World	28 yo. 1976	48.3%
2005	Jack Wiebers Whisky World	27 yo. 1978	46.7%
2005	Old Malt Cask (Douglas Laing)	27 yo. 1978	50.0%
2006	Gordon & MacPhail	1986	40%
2006	Limburg Whisky Fair	40 yo. 1965	46.7%
2006	Limburg Whisky Fair—Artist Edition	27 yo. 1978	56.8%

Third-party bottlings so far recorded

Date Bottled	Bottler	Age or Date of Distillation	abv
2006	Creative Whisky Company for Dansk Maltwhisky Akademi	20 yo. 1986	53.1%
2006	Creative Whisky Company	22 yo. 1984	52.2%
2006	The Vintage Malt Whisky Co.	29 yo. 1976	48.9%
2006	Cooper's Choice	29 yo. 1976	48.9%
2006	Milroy's of Soho	29 yo. 1976	46%
2006	Jack Wiebers Whisky World	20 yo. 1986	54.6%
2006	Douglas Laing for Parkers Whisky	30 yo. 1975	45.6%
2006	Exclusive Malts	22 yo. 1984	52.2%
2006	Murray McDavid	20 yo. 1986	55.3%
2006	Wilson & Morgan	23 yo. 1984	46%
2006	Signatory	38 yo. 1967	59.3%
2007	Signatory Vintage	30 yo. 1976	45.3%
2007	Cadenhead	23 yo. 1984	46%
2007	Cadenhead	23 yo. 1984	52.5%
2007	Berry Bros & Rudd	1983	46%
2008	Scotch Single Malt Circle	1984	54.3%
2008	Speciality Drinks Ltd.	30 yo. 1978	49.8%
2008	Dewar Rattray	34 yo. 1973	52.0%
2008	Single Malts of Scotland	30 yo. 1978	49.8%
2009	Planet of the Grapes	1984	54.2%

CHAPTER 4

Renaissance

On Tuesday 16th May 2006 Stuart Nickerson received a 'phone call which, unbeknown to him, would change his life. The caller, at that stage a stranger, was a senior executive in charge of acquisitions and diversification for The Scaent Group, a Dutch-registered investment company. He had an unusual mission: he wanted to buy a distillery.

Scaent's interests lie mainly in energy trading in Scandinavia and the independent states of the former Soviet Union. Founded as recently as 2003, Scaent has expanded to include interests in fields as diverse as property development, IT, public services and utilities, construction, retail, publishing and telecommunications with more than a dozen companies and sub-groups conducting business in 20 countries and across two continents.

The growth in global whisky sales and especially the dynamism of the luxury sector in Scandinavia, Russia and related markets had first caught their attention and then captured their imagination. The opportunity appeared a tantalising one, offering the company exposure to consumer markets, further developing its global reach but requiring long-term thinking. Could a distillery be acquired, they wondered, and a brand be built? The call to Nickerson was the result.

At this point, Stuart Nickerson was working as a consultant to the global distilling industry. Based on his many years experience, with spells working for Arthur Bell & Sons, Diageo, Highland Distillers and latterly William Grant & Sons, he had built up an expertise in production trouble-shooting. His consultancy work took him around Scotland and to the Caribbean, advising on distillery openings, refurbishments, energy usage and environmental best practice. It was a comfortable, interesting and, at times, reasonably lucrative niche (he might disagree with the 'lucrative'). Though far from the lifestyle of a corporate high-flyer it suited Nickerson's independent caste of mind and love of travel.

Scaent were clear about what they required in a distillery: high-quality spirit above all, an interesting and authentic heritage and, if possible, aged stock with which to re-enter the market. With these criteria the idea of building a brand new distillery was quickly considered, then abandoned: though the spirit quality could be assured through careful design and by utilising Nickerson's considerable distilling expertise, a start-up could never have the required tradition and patina of history and, by definition, aged stock is an impossible dream for a new-build site.

So began a lengthy search. Shortlists were drawn up; anonymous visits undertaken and discrete enquiries made. Scaent were meticulous and demanding in their requirements—more than one well-regarded name appeared for consideration and was then scratched as unsuitable. Nickerson's consulting work began to take a back seat as his new client increasingly dominated his time.

By May 2007 two separate deals had progressed significantly; in both cases to the point where a handshake had been exchanged and for one purchase, contracts drafted. But both fell through, due in both cases I am assured, to the potential vendor

The old malting floor

The old stone steep

withdrawing from the negotiation. Knowing the locations involved, but sworn to maintain strict confidentiality I may still observe that this is truly a case of third time lucky: *Glenglassaugh* proved a happy, indeed an inspired choice.

So far as anyone knew, *Glenglassaugh* was not for sale. It was certainly not actively marketed and, for the most part, had largely been written out of whisky's story. Its owners had sold other small distilleries as surplus to their requirements, but these were operating plants and came with a workforce, brand reputation and filling contracts. *Glenglassaugh*, then mothballed for some 22 years, was a very different proposition.

However, following the breakdown of the second set of negotiations to purchase another site, Nickerson remembered his days in Portsoy and got in touch with Graham Hutcheon of Highland Distillers, an old industry contact. Might *Glenglassaugh* be for sale, he enquired.

Initial discussions were positive. Highland Distillers proved amenable to the idea of selling; realistic about the price and prepared to make available all the aged stock which they held. Matters proceeded rapidly, though curiously another potential buyer appeared on the scene during the transaction. The negotiation began in May 2007 and proceeded, under conditions of great secrecy, until 29th February 2008 when payment was made and the transaction closed. Stuart Nickerson, slightly stunned at the magnitude of what he had just done, received the keys to *Glenglassaugh*, distillery, warehouse, dilapidated office and all, and drove onto the site, alone, to consider his next move. Much work lay ahead.

Simultaneously, the new company vehicle The Glenglassaugh Distillery Company Ltd came into being, consciously echoing the original trading identity. Nickerson was appointed Managing Director (initially he was the sole employee) and a new chapter in *Glenglassaugh*'s history opened.

At this point in the story, a diversion. Having worked closely with Stuart for somewhat over a year I have been mightily impressed by the Scaent approach to their new baby. It might have been expected that a procession of executives and shareholders would arrive in Scotland, all with ideas and pet projects that demanded immediate execution. It would not have been unreasonable to have anticipated a high degree of day-to-day involvement or, less desirably, remote management by memo and e-mail. The appointment of new management or, worse fate, consultants would have been understandable though, all too often, they arrive with little or no knowledge of the industry.

In practice, none of this has happened. Scaent and their investors are largely anonymous, deliberately so. They have adopted a distant 'hands-off' posture, keeping out of the picture and preferring to trust the management to manage; taking the view that the current team know what they are doing. Of course, reports are prepared and reviewed; budgets are submitted and plans discussed but, in general, the regime is a benign one. This is not, it would appear, an acquisition driven by corporate egotism nor is it some kind of 'trophy' asset (as was thoughtlessly suggested in some trade press reports at the time). Scaent appear to understand that the distillation of Scotch whisky is a long-term business, requiring patience and a steady nerve through global booms and recessions. Where a case for investment has been made the investment, as we shall see, has followed. Money has not been foolishly or lightly spent; neither has any reasoned request been refused.

This modest, low-profile approach seems to me a wholly admirable even, dare one suggest, a traditionally Scottish philosophy and follows from the lengthy period of consideration given to this strategic diversification. Distilling whisky is a long way removed

from the no doubt exhilarating and volatile world of energy trading: the Scaent team have had the great good sense to understand there are some aspects of their new acquisition that they do not fully understand and, instead, have put their trust in those who do. It seems well placed, though only time will tell. As a long-term business perhaps distilling provides a counter-weight to the vagaries of the energy market.

One further example of their stewardship is telling, however. Shortly after the new company began to restore the distillery the question of membership of the Scotch Whisky Association (SWA) arose. Though the members of the SWA account for the production of more than 90% of all Scotch whisky not all distillers, in particular not all the small independent firms, are in membership. Some are happy to exist in parallel with the SWA, following its guidelines and policies and benefiting from at least some of its work, whilst avoiding the tiresome necessity of paying a subscription whilst others, it is only fair to record, do not agree with the SWA and, on occasion, vociferously object to its activities.

The subject was not long debated at *Glenglassaugh*. Anxious to demonstrate their commitment to the industry and to conform to its standards and norms, The Glenglassaugh Distillery Company became the SWA's 54th member company in July 2008.

Becoming members of the SWA was a conscious and considered step to furthering their ambitious plans. As Stuart Nickerson said at the time, *'At Glenglassaugh we believe that we have a unique opportunity to breathe life back into a hidden gem. The SWA is of vital importance to the whisky industry and we recognise the part they play in ensuring that brands like ours become part of Scotland's future whisky heritage.'*

For the SWA, spokesman David Williamson responded, *'That Scotch whisky is to be distilled again at Glenglassaugh is good news for the industry. We are delighted Stuart and his team are going to play their part at industry level and welcome them as the SWA's fifty-fourth member company.'*

This may all sound like a carefully orchestrated round of mutual admiration but there was no pressure on *Glenglassaugh* to join the SWA and no requirement for them to do so. The option to remain outside the fold was a real one and the obligations of membership do to some extent constrain the company's freedom of movement, so the decision to seek membership should not be under-estimated, especially considering the implications for proprietors new to the industry and its expected codes of behaviour. Scaent clearly wanted to 'join the club'.

But this is to jump ahead. Though a comprehensive technical survey of the plant had been made this had not, and could not, reveal all the challenges that faced the new owners.

One set of problems was particularly poignant.

Though *Glenglassaugh* had remained silent since 1986 the buildings had remained secure and all the equipment remained intact for more than twenty years. Then, on 12th September 2007 and at a delicately-poised stage of the negotiation to acquire the site, *Glenglassaugh* suffered a break-in. This was more than casual opportunism—around this time a number of silent distilleries across Scotland were the target of an apparently co-ordinated series of attacks by thieves in search of scrap metal. As Stuart Nickerson told me: *'I was alerted on my way travelling North and arrived to find a scene of devastation! Police were alerted and installed a temporary motion sensor detector within the distillery buildings but the intruders never returned.'*

Though eventually a gang was caught and prosecuted, this was for a single, similar offence at Imperial distillery. The perpetrators of the *Glenglassaugh* break-in have never been formally identified and the two incidents may well be entirely

unrelated. Whoever was responsible, the damage was considerable and set back the process of re-opening by several months, not to mention the cost incurred. To purloin a few thousand pounds of scrap metal, it's estimated that more than £100,000 of damage was done. While it was clear why some vessels and pipes were removed, some of the destruction seemed little more than casually vindictive vandalism.

Gentlemen to a fault, Highland Distillers adjusted their asking price to reflect the cost of repairs but the incident was a disturbing one that caused no little upset to all involved.

But this was only one of the problems facing Stuart Nickerson. At the start he was working on his own: clearly a team had to be built up. Here he was fortunate.

Discussions were quickly opened, and equally quickly concluded to the satisfaction of both parties, with Graham Eunson, a quietly-spoken Orcadian. In a varied career, he had experience at both Glendronach and Scapa distilleries before taking charge as manager at the renowned Glenmorangie distillery in Tain. Fortunately, Graham was looking for a new challenge. Having had the depressing experience of closing distilleries the appeal of re-opening one was too great to resist. He was appointed on 1st April 2008, apparently undaunted by the date. The announcement attracted some curiosity and interest in the trade and amongst whisky aficionados. As he told them:

'During my career I have had to oversee the closures of both Scapa and Glendronach distilleries. So the opportunity to breathe life back into a mothballed distillery of such iconic status was one I couldn't resist. Until now, I feel that I have very much been the custodian of existing brands. With Glenglassaugh, I have the unique opportunity to make my mark on the whisky industry with a exciting new product.'

At much the same time, I was fortunate enough to be asked to work with Stuart and Graham on some aspects of *Glenglassaugh*'s marketing and publicity. Just as with the distillery, there was much to do: there was no corporate or brand identity; no packaging; no distribution network; no established products; no website or literature and virtually no awareness of *Glenglassaugh* even among whisky enthusiasts, though the announcement of the sale and planned re-opening attracted some attention and comment.

Work was therefore put in hand on many fronts at once. Contractors were appointed to deal with the main refurbishment of the distillery; the well-known consultant Dr Jim Swan was engaged to assist in the assessment of the casks purchased from Highland and we began work on a corporate identity and brand positioning strategy. Stuart Nickerson's days passed in a blur of activity, phone calls, e-mails and decisions. From time to time, well-known whisky writers and key independent retailers were invited to the site; it was fascinating to see their reaction to a time capsule, closed for more than twenty years.

The timing was tight: a commitment had been given to the new owners that whisky would be on sale and the distillery in operation by Christmas 2008. Some observers were frankly sceptical of this target and I recall discussing the date with one of the principal contractors (who had now better remain anonymous, though his firm did deliver on time).

'*Can't happen. Won't happen*', was the blunt verdict. Walking onto the site in early March 2008 it was hard to disagree. On a blustery and slightly damp day *Glenglassaugh* presented a forlorn and sadly abandoned appearance. So it was decided to photograph it.

The logical person to turn to was Ian MacIlwain, a retired NHS consultant psychiatrist turned photographer whose great love of Scotland's distilleries joins a keen eye with a profound sense of the romantic. In his photography, he combines lyricism with nostalgia. Ian had already photographed the *Glenglassaugh* site as part of his continuing and slightly obsessive quest to capture the almost Gothic dereliction of Scotland's silent distilleries before they either crumbled to dust or were sanitised by the health and safety lobby. He accepted the demanding task with quite alarming alacrity.

As you will see from his photographs appearing here and throughout the book, *Glenglassaugh* had been maintained, but hardly cared for. However, it seemed both important and appropriate to capture the site from the day of the take-over through its transition to a fully working distillery once again. Hence, Ian visited and revisited *Glenglassaugh* over the following twelve months documenting the process of renaissance with methodical care and almost proprietorial zeal. The result is an archive at once unique and exhaustive that will, it is hoped, be of interest to future generations. It also helps the team measure, with quiet satisfaction, what has been achieved if ever they are daunted by what remains to be done.

The changes have been considerable. Whilst the main vessels and component parts of the distillery were in good condition the building itself was damp and required roof repairs. It also had to be re-wired completely and a new boiler installed—a process not without drama, when it was discovered that the proposed location for the chimney could not support the necessary foundations!

A new gas main was laid into the site; the old cottages converted to offices; the accumulated clutter and rubbish of many years was removed from many of the buildings, revealing their potential; the former Manager's house was adapted to provide a small hand bottling facility (it is the company's aim to contain as much activity on site as possible); the filling store was restored; everything was painted, sometimes twice.

While this was going on the key parts of the plant were undergoing thorough checks and restoration. Thanks to the intervention of the burglars, a new false

bottom was required for the delightfully antiquated and thoroughly traditional Porteus cast iron mash tun, itself something of a rarity. The thieves had attempted to make off with the mash tun's copper dome but its size and weight fortunately defeated them.

Elsewhere, the Porteus malt handling equipment and malt mill proved to be in sound condition and, after reconditioning, was soon running again. The mill itself is thought to be pre-World War II, the presence of bronze bushes rather than bearings suggesting this vintage. Behind the mill room the more modern malt bins were in good condition though the screw conveyor required a comprehensive overhaul and clean. An ugly structure on the roof designed to facilitate the removal of malt was itself removed, along with a tumble-down corrugated iron malt delivery bay.

Also swept away were the remains of the water softening equipment installed in the previous owners' attempt to emulate the Speyside character. The water at *Glenglassaugh* is notably hard and the new make is uncompromisingly Highland in style: happily, there is no intention on the part of the current team to attempt to force *Glenglassaugh* into uncomfortable conformity with an alien style; it is to remain true to itself. Graham Eunson is familiar with the distilling properties of hard water from his days at Glenmorangie and an enthusiast for its qualities, even if authenticity was not already a watchword for his Managing Director.

With that in mind, work was undertaken to secure part of the original Number 1 warehouse at the heart of the site. Though a section of roof had collapsed, it has proved possible to cordon this off (it will be restored in time) and safely use some of the dunnage space. A proportion of the newly racked spirit is therefore stored here, maturing exactly as it would have done in 1875 though with considerably greater experimentation with cask types.

Fortunately, the stills proved to be in good

condition. Though they too had suffered damage in the break-in this was largely superficial and, after a thorough inspection and testing by Forsyths of Rothes, they were once again commissioned—after intense and physically demanding deep cleaning. Forsyths also cleaned and re-commissioned the distillery's spirit safe, which had initially appeared beyond repair.

Finally, the four Douglas Fir washbacks proved to have been scrupulously maintained since the closure and following testing and cleaning they too were brought back into use (the two stainless steel washbacks are currently mothballed). In total, more than £1m was spent in nine months to bring a silent distillery back to life—a contrast, perhaps, to the various noisily trumpeted schemes to build new distilleries in various locations in Scotland. Most seem doomed to remain pipe dreams and the fantasy of over-enthusiastic promoters.

But a distillery is more than a series of large rooms filled with equipment. It needs a team to bring it alive and much time was spent during the summer of 2008 in recruiting suitably experienced and motivated staff to bring *Glenglassaugh* back to life, both operationally and administratively.

Once appointed, the new team had to be trained in the *Glenglassaugh* methods of working (Nickerson and Eunson had first to determine these) and then familiarise themselves with the equipment and its inevitable idiosyncrasies—there was no instruction manual and no-one familiar with the processes to guide the exploration. It felt like an adventure. A palpable sense of excitement and optimism was clearly evident.

During the autumn work progressed at an increasingly rapid pace. Decisions were taken on the initial products to be offered and the look and feel of packaging. New signage appeared at the distillery. Contractors' lorries made frequent appearances, delivering men and machinery. Media interest grew, as it became apparent that the renaissance of *Glenglassaugh*

was not a pipe-dream, or a vision, or a speculation but was, in fact, actively underway.

The date of 24th November 2008 was set for the official opening and the local MP agreed to perform the ceremony and make the appropriate remarks to an invited audience of VIPs, suppliers, local dignitaries, industry leaders and former employees. As will be clear from the Foreword, the 'local MP' is, in fact, Scotland's First Minister, Alex Salmond. Failure was not, therefore, an option: the distillery had to be ready.

Memorably it was, even if the paint was scarcely dry in some areas and the 'distilling' consisted of merely starting the mill and beginning the first mashing, albeit for the first time in twenty-two years. Despite a throat infection the First Minister spent several hours on site, toured the distillery, mingled with guests in a filling store unrecognisable (for a few hours at least) and reflected on the disappointment he had felt as a young Parliamentarian when a vital part of the community life of his constituency disappeared with *Glenglassaugh*'s closure. His speech was witty, non-partisan and effective. Guests were charmed.

By restarting the mill, Salmond brought *Glenglassaugh* alive once again, but the real test came a few days later.

I was anxious to be present on Thursday 4th December for the first run of new make spirit. This is a rare occasion, not to be missed, and I watched the carefully feigned nonchalance of my colleagues turn first from relief through delight as they first tentatively nosed, then tasted the 'clearic'. That they were confident cannot be doubted, and for that they had good reason but for all their experience, for all the investment and for all the testing this was a tense moment. But it lasted only a moment: the spirit nosed cleanly and was fresh, with a delightful grassy quality, holding the promise of fine and delicate single malt once the necessary maturation had taken place.

The first casks were filled a fortnight later, the global financial climate having altered radically in the preceding months. The relentless optimism of the summer seemed a world apart from the gloom of the winter of 2008 and the following spring.

But the promise made to the owners and investors had been met. Whisky was being distilled and the first products were on sale by December. So *Glenglassaugh* pressed ahead. Tougher conditions had been weathered in the past and no one doubted that whisky's fortunes would change over the years ahead. Scaent's investment in purchasing and re-opening *Glenglassaugh*, said to be at least £5m and probably more, was not lightly made.

A sense of optimism, tempered with a realistic assessment of the challenges ahead, characterises the atmosphere at the distillery today. Whilst much has been achieved, much remains to be achieved not least with the various attractive but redundant buildings that remain on the site. The potential is clear though and, for the first time in many years, there seems a vision for *Glenglassaugh* that will see it eventually take its rightful place as one of Scotland's leading single malts (and if that seems like special pleading consider the outstanding achievement in the 2009 IWSC awards—by this measure at least, *Glenglassaugh*'s 40 Years Old expression could be said to be Scotland's finest whisky).

However, this is at least a ten year project. While some new products have been released (see the following chapter) these are mere striplings and it is on its mature spirit that the renaissance of *Glenglassaugh* will eventually be determined. A start has been made, an impressive start but only a start, as the current management and owners know full well.

It is clear that that single telephone call did not just alter Stuart Nickerson's life, but *Glenglassaugh*'s. Today, knowing them both, that seems both fitting and fated.

A changed life. Glenglassaugh at the Ronneby Whisky Society, Sweden, with Anders Bizzozero

A Note on Distilling at Glenglassaugh: The Facts & Figures

Malt intake and malt mill	Porteus—the mill is probably pre-World War II.
Malt variety	Optic and Optic/Oxbridge mix. Currently un-peated though limited amounts of a peated style will be distilled.
Mash Tun	Porteus. Cast iron, with rakes and copper dome. Diameter 15'.
Size of mash	5 tonnes.
Mash regime	Strike temperature 68.5°C, Mashing temperature 64.5°C, 2nd water approximately 80°C, 3rd water approximately 92°C. The worts are pumped to the washback at 18°C.
Washbacks	4 Douglas Fir washbacks, each with a capacity of 43,187 litres. The working capacity is 26,800 litres. The stainless steel washbacks are currently mothballed.
Fermentation time	Between 60–90 hours, depending when the washback was filled. Yeast—approx 70 kgs.
Distillation	A balanced system is operated, with a single pair of stills. One washback will fill the wash still twice; each charge of the wash still results in one charge to the spirit still. Therefore from one mash 2 runs of the wash still and 2 runs from the spirit still are obtained.
Still size	Wash still—17,200 litres; working capacity 13,400 litres. Spirit still—12,700 litres; working capacity 7,400 litres.
Still operation	The spirit comes into the safe at around 73% alcohol by volume (abv). Foreshots are run for 20 minutes before starting the middle cut; feints come on at 62.0% abv and the distillation is stopped at 1% abv.
Intermediate Spirit Receiver	11,260 litres. Located in the still room.
Spirit Receiver	37,304 litres. Located in the filling store. Filling strength 63.5% abv.
Distilling capacity	Approximately 22,000 litres of alcohol per week or some 1,000,000 litres annually.
Warehouse capacity	Dunnage: 1,000 casks mixed (currently, with room available). Racked: 29,500 casks mixed
Wood regime	A wide variety of casks are currently employed as the distillery experiments with maturation rates and the impact on spirit character.
Bottling	All bottling is now done on site. There is no chill filtration, no added caramel or colour and most products are bottled at cask strength (new spirit at 50% abv).
Staff	Office: 4. Distillery & Warehouses: 5.

CHAPTER 5

Something Old, Something New—The Current Expressions

Aspiring distillers of whisky, especially single malt whisky, require three qualities: patience, deep pockets and no little entrepreneurial spirit. For there is a problem with single malt whisky that does not trouble the distiller of vodka or gin or a number of other lesser spirits—the considerable period of time which must elapse before it is deemed ready for sale.

Forget the trifling legal minimum of three years aging: the distiller seeking to launch a single malt today at, say, twelve years of age (and some would consider this merely callow youth) would have had to produce this spirit in 1997 and then known the evenings, mornings, afternoons measuring out its life not with coffee spoons but a valinch.

It may not occur to the average distiller but there is a certain dignity, even a quiet courage, in that unhurried calm. In general, malt whisky distilleries are peaceful places. Not through any lack of activity, but from the knowledge that time does much of the work and that nothing in man's art or wit can replace that, or hasten it by even one day.

Glenglassaugh's new owners were faced with a particularly thorny version of this problem. Along with the distillery they had purchased Edrington's remaining stock but, because the distillery had been silent since 1986, there was a gap of more than twenty years in the age of the available whisky.

Thus the current expressions fall into two distinct camps which, not particularly imaginatively, may be characterised as the old and the new.

The old whisky is, in general, rather fine. This is not simply my partisan opinion. Stuart Nickerson undertook a careful evaluation of the stock prior to purchase, assisted by the highly regarded independent whisky consultant Dr Jim Swan (he advised Kilchoman on spirit development and wood policy, for example, and his expertise is sought internationally). This was a major consideration in the decision to press ahead with the purchase, as both concurred that the casks that remained varied from very good to excellent, with very few exceptions.

In this much credit must be given to *Glenglassaugh*'s former Managers (listed on page 38) and to the enlightened and far-sighted wood policy adopted by

Dr Jim Swan nosing a sample

Stuart Nickerson receives the IWSC Whyte & Mackay Trophy from Richard Paterson (left) and Sir Ian Good (right)

Robertson & Baxter in particular. That the spirit has aged so well and so delicately for so long is a tribute to the care exercised in its original distillation and storage. Accordingly, today the management at *Glenglassaugh* has the luxury of choosing from remarkably high quality mature stock.

Three aged expressions are currently offered. The decision was taken to trade on the exclusivity and quality of the whisky and to use memorable anniversary ages in the bottling, rather than bottle individual casks. This policy could, of course, change over time but the present policy is to offer 21, 30 and 40 year old styles.

In every case, however, the whisky is a little older than the nominal age—it will be evident that, not having distilled since 1986 the youngest whisky in store must be at least 23 years old but 21 years old was considered an age declaration more attractive for marketing purposes, and so on.

Stock is also being maintained with a view to an eventual 50 year old expression; this being anticipated for 2013 at the earliest. Other whiskies are being considered for release in the next few years probably based on single cask bottlings, perhaps in association with leading independent retailers and possibly commemorating significant individuals associated with the distillery or major events in its history.

It was not long before the quality of the product was recognised by external commentators. Both the 30 and 40 year old were entered by Stuart Nickerson in the relevant 2009 International Wine & Spirit Competition categories in the confident expectation of at least a Gold medal.

His hopes were hugely exceeded: both received Gold medals but beyond this, both were declared 'Best in Class' and, remarkably, both were awarded the prestigious Trophies on offer. The 30 year old was declared the Whyte & Mackay Trophy winner for Cask

An Octave owner signs his cask

Strength Scotch Whisky and the 40 year old collected what may be considered the blue riband trophy, the IWSC's own Trophy for 40 year old Scotch Whisky. This was a special trophy instituted in 2009 to mark the IWSC's own 40th birthday and, as such, especially esteemed and highly coveted.

In the pages which follow, the IWSC judges' tasting notes for these products are given in full, a vindication of the judgement made by Nickerson and Swan a year earlier when first sampling casks of these venerable whiskies.

But such products are necessarily expensive and in restricted supply. *Glenglassaugh* needed to get its name more widely known and offer a more accessible product.

Clearly this could not be marketed as 'whisky' as it could not fulfil the legal requirement of three years aging, so the decision was made to bottle 'new make' as *The Spirit Drink that dare not speak its name*. In the interests of objectivity and balance, I must declare an interest here for I was closely involved in the development of this product.

'Spirit Drink' is the appropriate product classification for distilled spirits which do not fit into the other categories defined in European Union law. It seemed rather a blunt, even harsh descriptor but, as we worked with the idea, we saw that it was possible to treat it with some wit and flair. Hence the tongue in cheek name and labelling, which describes the product, entirely honestly, as follows:

> BE INFORMED: *this is not Single Malt Scotch Whisky. If for three solitary years we kept it in oak casks in our coastal Highland warehouse it would be, but we simply couldn't wait. This is Gleng------gh's New Spirit with all the joy, hope and glamour of life before it, a spirit that is as pure as it is perfect. It is beautiful, it is fine, it is good enough to drink.*

The name, of course, is a reference to Lord Alfred Douglas' poem 'Two Loves' which brought such difficulty into the life of Oscar Wilde, and the copy on the front label is, I can now admit, shamelessly plagiarised from Wilde's defence of his relationship with Lord Alfred, the aim being to associate some clandestine glamour with a product otherwise hampered by a rather prosaic appellation. The suggestion of something faintly covert or secret, if not illegal, we found an amusing nod to the product's flaunting of Scotch whisky convention but we reconciled ourselves to that with the thought that, in all probability, this is how much of the very first *Glenglassaugh* was drunk.

Something else unusual about *The Spirit Drink that dare not speak its name* is that it is the product of a single mash. Looking for a legitimate way to limit the initial release we hit on the idea of turning normal practice on its head and releasing a single mash rather than a single cask. The result: 8,160 bottles at 50%

abv—probably a world first, offering the connoisseur and enthusiast a different insight into distilling practice and flavour development.

Certainly Gavin Smith of whisky-pages.co.uk was impressed, immediately declaring it his 'Whisky of the Month' for July 2009, though having promptly to explain that it wasn't actually whisky. He found it *'offers a slightly cereal-y nose of cream and summer meadows, while water teases out fresh fruits. Initially peppery and 'upfront' on the palate, pleasing, brittle toffee notes emerge with time, while dilution leaves an extremely drinkable, smooth, fruity dram with spicy overtones. The medium-length finish is sweet and spicy.'*

A remarkable variation quickly followed. Radical and innovative experimentation with the impact of different cask types brought *The Spirit Drink that blushes to speak its name*. Launched in July 2009, this is new spirit matured for just six months in ex-wine casks from California, a procedure generally avoided because of the risk of rapidly over-powering the initial spirit character. However, Nickerson and Eunson watched over these casks very carefully, taking and comparing samples on a weekly basis and their careful attention has paid off with a spirit as unusual as it is exciting.

The rosé blush hue is immediately noticeable and the impact of the aging dramatic: an initial aroma of yeast is followed by intense red berry fruitiness and a hint of spicy honey. One of the benefits of *Glenglassaugh*'s small size is the ability to undertake this kind of bold experimentation: a visit to the Number 1 warehouse will confirm that there are other surprises in store!

Glenglassaugh also offers consumers the opportunity to purchase their own cask of new make spirit and choose when to bottle it. A considerable range of cask sizes and finishes are available, ranging from an Octave (50 litres approximately, and expected to mature in 5 years or less) up to a Butt (500 litres). The larger casks

may, one day, produce whisky of IWSC Gold medal standard; certainly that is their provenance.

The cask sale programme has been enthusiastically taken up with, at the time of writing, owners in the UK, USA, Japan, Switzerland, Germany, Holland, Sweden and a number of other countries. The farsighted purchasers of the larger casks obtain membership of the *Glenglassaugh* 250 Club offering a range of privileges at the distillery.

In the pages which follow the distillery's current products are detailed, with tasting notes and further background. In the interests of objectivity, independent tasting notes have been used where possible but I have also added my own comments and observations.

OCTAVE CASKS

The Octave is a cask of approximately 50 litres (11 gallons, and thus slightly larger than the traditional firkin) the use of which *Glenglassaugh* have revived primarily for private sales.

It has three advantages: being smaller it matures faster, costs less and offers a volume of finished whisky more suitable to the individual. Having been largely abandoned by the whisky industry, *Glenglassaugh*'s offer of single casks has been enthusiastically taken up by knowledgeable consumers.

The distillery are themselves aging spirit in their own Octaves with a view to an early release of single malt and experimenting with a range of unpeated and peated styles.

250 CLUB CASKS

The 250 Club is for private sales of casks in larger sizes than Octaves and with a greater choice of wood types. *Glenglassaugh* have sourced first fill and refills from ex-Bourbon, ex-Sherry, ex-Port, dechar/re-char and ex-wine casks from many regions, including world-famous Bordeaux chateaux and fine red wines from California and are also carrying out their own experimentation on the impact of wood on the distillery's new spirit.

Cask owners in the 250 Club programme will have a range of privileges and facilities at the distillery and the intention is to create the feel of an exclusive 'private members' club.

THE SPIRIT THAT DARE NOT SPEAK ITS NAME

This is new make spirit—the raw material that will, with time, become single malt whisky. Bottled at 50% abv this has been enthusiastically received not simply as a curiosity but as an indicator of the likely quality of the eventual whisky. 'Recommended' by F. Paul Pacult and selected as 'Whisky of the Month' by whisky-pages.com.

Colour: Clear and colourless, like mineral water.
Nose: Tropical fruits and flowers; pears, green apples and citrus fruit.
Taste: Fruity, slightly sweet; very clean. Floral.
Finish: Fruit notes, some spice hints.

THE SPIRIT THAT BLUSHES TO SPEAK ITS NAME

New spirit is left for approximately six months in ex-red wine casks and matures dramatically, taking on a rose blush. This has been enthusiastically adopted by a number of fashionable leading-edge cocktail bars, most notably Edinburgh's Bramble Bar (voted one of the Top 20 Bars in the World by *Bartender magazine*) and was 'Highly Recommended' in F. Paul Pacult's Spirit Journal.

Colour: A striking rose-pink.
Nose: Wine notes, with fruit character developing.
Taste: Ripe fruits, melon and William pear; grapey with late spice development.
Finish: Lingering fruit and red wine notes.

21 YEAR OLD

Arguably *Glenglassaugh*'s 'entry-level' expression (but a very serious whisky), just 8,700 bottles were released—and when it's gone, it's gone, reflecting the shortage of aged stock. Handsomely and distinctively presented in a 'pear drop' shaped bottle, F. Paul Pacult said it 'definitely deserves a home in the US' and Jim Murray found it 'elegant and adroit'; concluding it was 'a fabulous benchmark for the new owners to achieve in 2030!'

I find it both intriguing and surprising: clearly mature, yet still fresh, delicate and complex. The 21 Years Old (actually it's somewhat older) is beguiling and alarmingly easy to drink, rewarding careful consideration whilst never petulantly demanding your attention.

Colour: Pale gold.
Nose: Surprisingly delicate; dried fruits; noses younger than the age would suggest.
Taste: Complex and multi-layered; spice notes dance with rich fruits and caramel; superb balance.
Finish: Vanilla notes carry freshness through into a long finish.

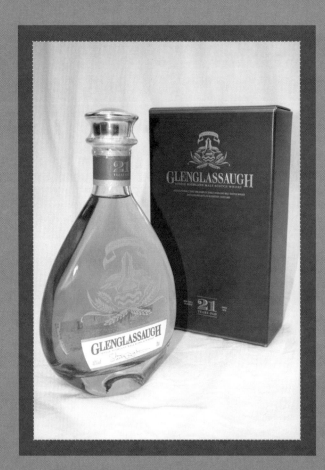

30 YEAR OLD

A major trophy winner in the 2009 IWSC competition, this whisky is selected from single casks and has attracted high praise for its lightness, combined with graceful aging.

Jim Murray found it 'sheer poetry' and the IWSC judges said: 'Exceptionally complex nose where citrus plays an important part and a gentle note of peat lingers in the background along with some floral notes. Deep mouth is redolent with maturity and layered flavours of marmalade, dark chocolate, peat and sweet spice. Slow, full, smooth across the palate. Velvet texture covers a firm body. Long, attention-retaining finish.'

Colour: rich gold.
Nose: Rich and creamy toffee and raisin fruit cake aromas are quickly followed by a stunning array of fruit notes. Roasted almonds in the background.
Taste: an intensely sweet and powerful whisky reminiscent of wedding cake opening up to a long, long finish. Hints of dark chocolate studded with orange peel.
Finish: Shows great maturity and continued spice development and complexity.

40 YEAR OLD

Again, a winner of a major IWSC trophy, the flagship *Glenglassaugh* expression and a touch of true luxury. Awarding it 96 points, Jim Murray described this as '…truly beyond belief. One of the great world whiskies for 2010'.

The IWSC citation reads: 'Citrus and oak dominate the nose with a fine floral backing. Deep, rich mouth has mellow mature feel with loads of marmalade and dark, bitter chocolate. Also some espresso notes. Touch of peat and fine smoky note. Smooth palate with slight dryness of old oak. Long, fine, firm finish.'

This is a characterful whisky perhaps best suited to after dinner sipping.

Nose: Strong and powerful, beginning with green apples falling to earth onto wet moss and grass. This gives way to a medley of boiled fruit sweets, and then hints of raspberries, blackberries and ripe banana emerge.
Taste: rich sherry notes of sultanas, fruit cake and cream fudge—an incredibly powerful yet complex whisky.
Finish: The finish is long but despite great age leaves the freshness of Granny Smiths apples on the palate.

CHAPTER 6

A Run Through Glenglassaugh With Alfred Barnard

Who was Alfred Barnard? Why did he visit Glenglassaugh? And why is this important? ¶ Alfred Barnard (1837–1918) is today revered as the father of whisky writing. Relatively little is known of his early life though by 1859, aged 22, he was married and working in London as a draper. ¶ By 1871 he had opened an advertising agency in Fleet Street. This venture did not last long however: by 1885, he was employed by the leading wine and spirit trade magazine Harper's Weekly Gazette who commissioned him to tour the United Kingdom and file a series of reports on the country's whisky distilleries.

He threw himself into this work with great energy. Scotch whisky was at that time undergoing both great changes and enjoying a spectacular boom. The advent of blending, made possible by the invention of the continuous still, had led to the formation of the great blending houses (Dewar, Walker, Buchanan, Haig and others) who were beginning to enjoy conspicuous success, just as the Irish whiskey industry began its long decline. Global sales were increasing; technology was beginning to make a significant impact on all industrial activity (not just distilling) and, with the growth of the British Empire it was a time of growth, expansion and confidence.

Barnard sent regular reports back to London to be published in *Harper's*. We can speculate that they were well received because, in 1887, they were collected into his first great work *The Whisky Distilleries of the United Kingdom*. *Glenglassaugh* received a modest entry in this volume.

Today, copies of the rare and valuable original first edition represent the 'Holy Grail' of whisky book collecting. As a record of the Victorian whisky distilling industry they are invaluable, though Barnard, conscious no doubt of his audience, devotes much space to extraneous detail of his travels and the landscape through which he passed rather than giving us, as we would wish today, very much technical detail on the distilleries themselves.

For all that, his work provides the definitive account of the distillation of whisky at a critical point in the development of the modern industry and it is keenly studied. Following the publication of his magnum opus he returned to work full time for *Harper's* and, between 1889 and 1891 published a four volume work *Noted Breweries of Great Britain & Ireland*, which enjoyed considerable success.

It should be understood, however, that Barnard was a very commercial writer and not averse to accepting commissioned work. Inclusion in his *Noted Breweries* was conditional, not just on the scale and prestige of the firm concerned, but on payment of a fee of 25 guineas to Barnard himself (around £2,160 in today's money based on the RPI). In addition, reprints of the individual entries were authorised.

By 1898 the opportunity arose to exploit his earlier work on whisky in a similar manner. The Highland Distilleries Company of Glasgow, formed in 1887, commissioned Barnard to visit their four distilleries (Bunnahabhain on Islay; Glen Rothes and Tamdhu in Speyside and *Glenglassaugh*) and write an account of this in a brand new pamphlet, reflecting the success and progress of the firm.

This had the engagingly romantic title of *A Run through some famous Scotch Distilleries*. The title page is decorated by vignettes of the four distilleries and superimposed above the title is a line from Robert Burns—'Willie Brew'd A Peck O' Maut'.

Similar commissions followed from John Walker and Sons; Mackie & Co; Watson's of Dundee; Dalmore distillery and the ill-fated Pattison, Elder & Co. All are now exceptionally rare and, in some cases, may survive only in single copies. *A Run through some famous Scotch Distilleries* has never been reproduced in facsimile. Chapter 4, which describes *Glenglassaugh*, is therefore shown here for the first time in more than 100 years and is a great rarity.

As a commissioned work, produced to gratify a paying client and promote their interests at a time of great competition, we need not look to Barnard's little pamphlet for disinterested commentary. Nevertheless, it is a fascinating record of *Glenglassaugh* some twenty-three years after it was built and the only known near-contemporary photographic record of the distillery after its 1887–1892 refit by Alexander Morrison. With its tantalising view of the interior of the still room and panoramas of the original distillery buildings it provides much of interest to the historian.

Alfred Barnard

Perhaps with an eye to pleasing his client Barnard described *Glenglassaugh* as *'too well known to need any praise'*. That may have been extravagant yet his final words ring true: *'though not one of the largest of the Company's Distilleries, yet one of the most interesting, distinguished alike by its cleanliness and thorough completeness in every detail'*.

Even today that would be considered high praise.

GLENGLASSAUGH DISTILLERY, BANFFSHIRE.

GLENGLASSAUGH DISTILLERY.

". . . There is rapture on the lonely shore,
There is Society where none intrudes,
By the deep sea, and music in its roar."
                                        BYRON.

CHAPTER IV.

LEAVING Elgin by an early train, a ride of one hour brought us to the pleasantly situated distillery of Glenglassaugh, the very ideal of seclusion and repose. Travelling thither by the coast railway, we were interested in viewing—as we passed along—the diversified face of the country, broken as it is, into hill and dale, with every now and then rich wooded bounds, enclosing large tracts of corn lands. After leaving Fochabers the eye takes in one of those landscapes so frequently to be observed on the East Coast—a stream of water running to the sea, some level country, and a line of trees on rising ground leading up to mountainous hills; the latter breaking the monotony of the plain and giving an air of beauty to the distance.

Presently we rolled along the coast, and crossing a viaduct, reached the thriving seaport of Buckie, whose picturesque and busy harbour is seen below the lofty railway track.

> "Where the busy fishers gaily launch
> And trim their mimic fleet,
> And the billows on the pebbled shore
> Make music in their beat."

VIEW OF UPPER RESERVOIR, GLENGLASSAUGH.

Soon after this we reached the Glassaugh Station, from whence could be seen the roofs and chimney stack of the distillery we were bound for. Instead of taking the field path leading direct to Glenglassaugh, we diverged to the right, and made our way along a road leading to a bridge (a quarter of a mile distant), to take a glance from that elevation of the glen and burn beneath it, from which the distillery takes its name. Presently we descended to the banks of the stream, from whence, after skirting a clump of trees, we came to a mossy rill, creeping almost unseen, among tangling brambles and ferns and betrayed only by the surrounding verdure, whilst ever and anon as we pursued our way, we see it again, as it joins the swiftly flowing burn on its way past the distillery to the sea.

Higher up the glen in an isolated spot is the "Rumbling" Pot, so named from the fact that the waters of the burn tumbling over the boulders have worn out a deep excavation in the solid rock in the shape of a pot.

Retracing our steps to the bridge, we followed the course of the burn as far as the reservoir, where we beheld the Glenglassaugh Distillery, erected on the slopes of the hill and extending almost down to the sea.

WATERFALL, GLENGLASSAUGH.

At this point of our observations we were met by the Manager, who, apprised of our coming, was on his way to the station to meet us.

At his suggestion, and having plenty of time to spare, we mounted the rocky hill which rises at the foot of the glen and protects the premises from the inroads of the sea, to take stock of our surroundings.

Crossing a rustic bridge over the Glassaugh burn, we ascended by a winding path to the top of the acclivity, whence a fine seascape and a long extent of country is to be seen.

The first object that greets the eye is the celebrated Red Rock, which rises 100 feet almost perpendicular from the sea, and is only separated from our standpoint by a gap of a few hundred yards.

"There is a cliff, whose high and bending head
Looks fearfully on the confined deep."

LOWER RESERVOIR, GLENGLASSAUGH.

The sea, which is so frequently rough on this coast, beats furiously upon these boundaries of the distillery with an everlasting refrain, chanting its endless hymn to Nature.

Below the gap referred to was formerly the old harbour of Dinnichigh, which, during a furious storm at the end of the last century was completely swept away.

From this height the sea and coast views are very fine, made all the grander by the broad surf of waves that can be traced for many miles along the coast.

> " Oh ! wild the rush of waters, and the sound
> Of bursting breakers, and the moan of waves
> As they, in angry freedom, range around
> And dash upon the rocks, their white spray laves."

Round the point northward are seen the ruins of Findlater Castle, a miniature Gibraltar, which stands on a peninsulated rock overhanging the sea, and is a picturesque and curious ruin. It was formerly a place of considerable strength, and made some figure in the history of the feudal wars. This was one of the castles which refused to receive Queen Mary on her way to the north.

Looking down from the cliff above upon the straggling ruins of this castle, the mind is carried back to the ages

> " When might was right, possession law,
> In donjon keep, and castle ha'."

Having viewed with an admiring eye the bold scenery of this rugged coast, we next proceeded to make a tour of the establishment so intimately connected with the object of our visit.

The Glenglassaugh Distillery is situated in the parish of Fordyce, Banffshire, on the banks of the Glassaugh burn, and was erected in the year 1875 on a site of a smugglers' bothy. Owing to

GLENGLASSAUGH DISTILLERY WATER HEAD.

its close proximity to the sea, the Glassaugh glen was in former times the home of several illicit stills, and considerable smuggling operations were carried on there. This was owing to its favourable position on the coast, as the smugglers were able to send their whisky by boat to Aberdeen and Leith whenever they could decoy the Revenue Officers to Portsoy or elsewhere. The distillery is about ten minutes' walk from the railway station and is situated in a fine corn-growing district, and splendid barley crops are gathered almost at its very doors, a boon which the Company freely avail themselves of.

EAST SIDE, GLENGLASSAUGH.

WEST SIDE, GLENGLASSAUGH DISTILLERY.

In close proximity to the works, in a field adjoining the distillery, stands an ancient mill building which ages ago also did duty as a watch tower. It is a stone structure of considerable proportions, and from the summit fine views of sea and land can be obtained.

The distillery comprises a group of solidly constructed buildings erected on the slopes of a steep hill; hence all the processes are conducted by gravitation, and water is the only power used in the works. Although built so many years ago, the establishment has been thoroughly renewed and equipped with modern appliances and vessels, of the same modern type as those in the Company's other distilleries already described.

We were conducted over the premises by the "brewer," who took us first to the granaries. One of these, a building 123 feet in length, contained at the time of our visit a vast store of golden grain; a considerable quantity of barley is also kept in the top story of the maltings. These latter are three stories high, and possess the usual cisterns, elevators and machinery. Passing the kilns we were shown the malt storehouse, a new detached building, which is in communication with the mill-house by means of an elevator and screw.

Walking a few steps up the hill, we entered the principal building of the group, nearly 90 feet in length, which contains the mash-house, fermenting and distilling departments.

It is not our intention to enlarge fully upon the arrangements and paraphernalia of this well-known and famous distillery, suffice it for our purpose to notice a few salient points. The cast-iron mash-tun, which is commanded by a Steel's mashing machine, possesses an ingenious invention for sparging the grains after the worts have been drained off into the underback. From the latter vessel the worts are pumped (the only pumps on the premises) through a copper pipe 90 feet in length to a vessel in the cooling department.

Here also is a Miller's refrigerator, where the worts are further cooled over the mazy lines of innumerable cold pipes before reaching the fermenting tuns.

The tun-room to which we next proceeded is a spacious apartment, containing five wash-backs, and the still-house, a modern structure, immediately adjoins it. The last mentioned

GLASSAUGH BAY, TO THE WEST OF THE DISTILLERY.

STILL HOUSE, GLENGLASSAUGH.

contains two pot-stills and all their subsidiary vessels, which are of the newest and up-to-date character. On the terrace below, situated on the banks of the stream, stand the spirit-store and two duty-free warehouses; there are also two others on the higher slopes, one of which is 160 feet in length and holds 3,000 casks. Adjacent to the warehouse are cask sheds, a cooperage, carpenters' shop and other places; whilst beyond these are several modern houses for the manager, brewer, and excise officers, also cottages for the work-people. The water supplying the worm-tub comes from the dam, brought thither by a conduit at the back of the distillery, the overflow being carried over an immense water-wheel which drives all the machinery in the place. Westward of the water-wheel are the peat stores, containing upwards of 500 tons of that valuable fuel; brought from the Crombie moss.

Glenglassaugh whisky is too well known to need any praise here; we may however, remark that it is pure Highland malt, and the annual output is over 100,000 gallons. As we made our

PHOTO. OF "DAVIE,"
A FAVOURITE OLD HORSE, AT THE
ADVANCED AGE OF 36 YEARS.

tour of the distillery, we were much impressed by the orderly manner in which the operations are carried on, and were surprised at seeing so many aged men at work; on enquiry we found they had been employed in the works for a great number of years, and in several instances, the son had succeeded the father, as was the case of the young brewer, our courteous guide.

Before completing our task we took a peep at the stables, where are stalled some capital dray horses, used for carting the whisky to the railway station. We were shown the photograph—here reproduced—of "Davie," a favourite old horse of Irish extraction, who died last spring at the unusually advanced age of 36 years. We cannot conclude this chapter without mentioning that the progress of the Highland Distilleries Company's business at all their distilleries has been materially assisted by the development of the English wholesale trade, and the formation of this interesting group into a limited company. Altogether this is one of the most thriving enterprises in Scotland, and its rapid success is entirely due to its careful management. The annual output of the whole list is fully one million gallons, so that this Company may be said to be one of the largest, if not actually the largest, distillers of Highland malt whisky in Scotland.

So good-bye to Glenglassaugh—though not one of the largest of the Company's Distilleries, yet one of the most interesting, distinguished alike by its cleanliness and thorough completeness in every detail.

> "But I must bid these pleasing scenes adieu;
> Farewell ye grazing herds, and warbling birds.
> I go to seek the savage haunts of men."

VIEW OF OLD WINDMILL AT GLENGLASSAUGH DISTILLERY

Credits & Acknowledgements

This book could not have been completed without the generous help of a number of individuals. In particular, I would like to thank Stuart Nickerson, Jim Cryle, Fraser Morrison, Professor Michael Moss and, for his kind foreword, the Rt Hon Alex Salmond, MP. Jules Akel, who designed the book, has been immensely patient with my constant revisions and changes and Stuart Nickerson assisted greatly in correcting my understanding of distilling. Any errors or omissions remaining in the text are mine alone. IAN BUXTON

PHOTOGRAPHY

Images for the following pages were kindly provided by:

Ian MacIlwain—*panoramic frontispiece, title page, x, 4, 9, 35, 59, 62, 63, 66, 67, 70, 71, 75, 78, 79, 83, 86.* John Paul—*ii*
Ann Smith—*vii, 92, 93, 95, 97, 98, 99.* Russell Cheyne—*53, 54, 57, 58.* Patrik Pålsson—*82*
Nick Lawson, *courtesy of the IWSC—87.* Katy Plume—*117, 118*

Postcard, page 2 and photograph, page 30 courtesy of John Mitchell
Photograph of Dugald Mathieson, page 8, courtesy of Mrs Margaret Turner
Postcard, page 12, courtesy of Jim & Linda Brown
Photograph, page 19; Pattison's advert, page 22; postcard, page 29—Private Collection
Photograph, page 27; aerial view, page 47, and Barnard pamphlet pages 103–114, courtesy of The Highland Distillers plc
Photographs, pages 32, 33 & 43, courtesy of Elaine Morrison
Photograph, page 42, courtesy of Jim Cryle
Photograph, page 102, courtesy of Andrew Barnard
All other photographs and illustrations, Glenglassaugh Distillery

ABOUT THE AUTHOR

Having worked with Glenglassaugh since March 2008 Ian Buxton knows the company very well. He is a former Group Marketing Director for a well-known single malt whisky. Today he writes widely on whisky, continues his consultancy work and is Conference Director of the World Whiskies Conference, which he established in 2006. ¶ His other publications include an introduction to 'Aeneas MacDonald's Whisky' (Canongate); contributions to 'Beer Hunter, Whisky Chaser'; 'Eyewitness Guide to Whisky' and 'World Whiskies' (both Dorling Kindersley) and 'Whiskey and Philosophy' (Wiley). He has written the company history of John Dewar and Sons, 'The Enduring Legacy of Dewar's' and numerous magazine articles in the UK, Europe and the USA. ¶ Ian is a Keeper of the Quaich, a Liveryman of the Worshipful Company of Distillers and a Grand Ordinary Member of the von Poser Society of Scotland.

Bibliography

ALFRED BARNARD
The Whisky Distilleries of the United Kingdom, Harper's Weekly Gazette, 1887
A Run through some famous Scotch Distilleries, The Highland Distilleries, 1898

CHARLES CRAIG
The Scotch Whisky Industry Record, Index Publishing, 1994

LAWRENCE GIPSON
The Great War for the Empire, Alfred A Knopf, 1965

RICHARD JOYNSON (ED)
Scotch Whisky Review, Issue 16, Loch Fyne Whiskies, 2001

S.R.H. JONES
Brand Building and Structural Change in the Scotch Whisky Industry, University of Dundee, 2002

AENEAS MACDONALD
Whisky, The Porpoise Press, 1930

CHARLES MACLEAN
The Robertson Trust, Maclean Dubois, 2001

MICHAEL MOSS
100 Years of Quality: Highland Distilleries 1887–1987, Unpublished MS, 1987

MICHAEL MOSS & JOHN R. HUME
The Making of Scotch Whisky, James & James, 1981